HOW MEMORY SHAPES NARRATIVES

A Philosophical Essay on Redeeming the Past

HOW MEMORY SHAPES NARRATIVES

A Philosophical Essay on Redeeming the Past

Theodore Plantinga

The Edwin Mellen Press
Lewiston/Queenston/Lampeter

Library of Congress Cataloging-in-Publication Data

This volume has been registered with The Library of Congress.

ISBN 0-7734-9575-4

A CIP catalog record for this book
is available from the British Library.

The Edwin Mellen Press
Box 450
Lewiston, New York
USA 14092

The Edwin Mellen Press
Box 67
Queenston, Ontario
CANADA L0S 1L0

The Edwin Mellen Press, Ltd.
Lampeter, Dyfed, Wales
UNITED KINGDOM SA48 7DY

Printed in the United States of America

Contents

Chapter 1

Introduction: The Elusive Past

Does the past exist? Many of the most interesting and difficult questions we encounter in human life (including philosophical questions) are of such a nature that one can plausibly answer them with either a yes or a no. One such question is whether the past exists. That something which we usually call "the past" can be referred to and that it often needs to be discussed would hardly be doubted by a rational person. Yet one need not be a philosopher to be aware that there is also a sense — perhaps puzzling, but also existentially relevant — in which the past is no more.

Some people, perhaps without ever engaging in philosophical reflection, do seem to believe that the past, or perhaps part of it, can be denied or said not to exist. They put their denial of the past into practice in a very down-to-earth way — by denying agreements and contracts they have made in the past, such as marriage. A man might tell his wife that he is not the same person who made those marriage vows twenty-five years ago — therefore he does not consider himself morally obliged to continue his relationship with her.[1] Or a shrewd and cynical manipulator may realize that when all the parties and witnesses to a sordid episode in his own life have died off, the episode

1. The man in this case is obviously a male human being, but in many other passages the words "man" and "mankind" will be used in a generic sense that includes both male and female. Pronouns referring to such instances of "man" are to be understood in the generic sense as well.

can be disregarded in terms of its impact on the present. For such a person, the events of yesterday and even yesteryear do continue to exist, but it may be that certain events of fifteen or forty or more years ago have, for all practical purposes, ceased to be real. Surely this is — or was — the mentality of many war criminals from World War II days who were never brought to justice.

"Forget about it," we sometimes say to one another when trying to come to terms with a distressing or unpleasant event. In many situations, especially when we are dealing with comparatively trivial matters, this is probably wise counsel. But Christian reflection reminds us that on a deeper level, the past must *not* be forgotten, ignored, blotted out. God chose *not* to look the other way when he was confronted with mankind's original rebellion against his decrees. And he undertook to set things right again in such a way that a chain of events (which Christian theology calls the history of redemption) *had* to take place and then be recalled throughout subsequent history. "If Christ has not been raised, then our preaching is in vain and your faith is in vain," writes Paul (I Corinthians 15:14).[2] Moreover, Christian worship presupposes the reality of the past and incorporates it. "Do this in remembrance of me," we read on many a communion table (see I Corinthians 11:24–5). When it comes to our original sin and Christ's redemptive death and resurrection, we may not "forget about it."

There is considerable theological discussion and disagreement regarding that first sin and also Christ's crucifixion and subsequent victory over death in the resurrection. From this tradition of reflection we see that even on a theological level, attempts have been made to minimize the reality of the past. Some theologians have supposed that whether there really was such a first sin or not makes no difference: the *story* of the first sin sheds light on the human situation in general. They also tell us that it does not matter whether Christ actually arose from his grave.

What is important for me as I seek to explicate and defend my thesis regarding our relation to our personal past is that Christian

2. Bible quotations are taken from the Revised Standard Version.

orthodoxy has insisted that mankind's original rebellion and Christ's resurrection are matters of the greatest significance. And while no *direct* line can be drawn from these theological themes to the questions we encounter in philosophy of history, I maintain that it is in the spirit of Christian thinking to insist that the past is real. To produce and retain in consciousness a correct, faithful account of the past is a sacred obligation. (We are to tell these things to our children: see Psalm 78, especially verses 1–4.) Therefore we may speak of *truth* in connection with history, our awareness of the past.

Can the past be changed? This question, too, must be answered with both a yes and a no. Common sense is at first inclined to offer only the no — what's done is done! Yet a little reflection reveals an everyday sense in which the past *can* be changed, for there are indeed some actions that can be undone. When furniture is rearranged and the effect is disappointing, it is usually possible to restore it to its original arrangement, and no one need be any the wiser. When a boy comes home from the barber with a botched haircut, we assure him, "Your hair will grow back in time, and you'll look just the way you did before."

There is a deeper sense in which the past is open to change. To grasp that sense, we must realize that the actions, events and episodes that make up human life are not atomic in character; that is to say, it is not the case that events and actions are self-contained, self-sufficient, and properly understandable and describable without reference to anything outside themselves. Philosophers sometimes speak of actions as analyzable into various subactions. Presumably the actions that are made up of subactions can in turn be regarded as subactions in relation to some yet larger action that is unified by a longer-range intention. Once this is realized, we must ponder an interesting possibility, namely, that what we do in the future may pull the actions we have undertaken in the past into some larger web or context in a way that alters them in their significance, which should in turn affect how we remember and retell them.

If we admit such a possibility, we are allowing the past to be changed on at least two levels. First of all, we are opening the possibility that future actions we can perform will change past actions. Secondly, we are allowing for a redescription of past actions in the light of future actions or perhaps — and here a new element enters the analysis — in the light of *possible* future actions. The latter are actions we have not yet undertaken, actions we are merely contemplating and might not even be in a position to perform for quite some time. We run ahead of ourselves in narrating, anticipating what we might do and intend to do, anticipating how our future actions will affect the way we later sum up what we have done and are doing.

Although people do not reflect on these matters very much, they do engage in this type of activity: they change the past in imagination, on the basis of intended action and of the narratives which such actions will permit. And while such changing of the past opens the door to much hypocrisy and self-deception (which is why so many people are wary of it), it also allows us to set right — if only in part — that which went wrong, perhaps because of unforeseen factors, and even that which we knowingly did wrong in the past.

The notion of changing the past (perhaps even redeeming it) is essential to the Christian life. Indeed, the Christian understanding of salvation is based upon it: God does not abandon or repudiate what he has made but chooses to bring good out of bad, and to redeem and restore what has gone astray. He holds open a similar possibility to us, albeit on a lesser scale. The Christian life of daily conversion is not a matter of forsaking the past; it is better understood as an effort to make something of the past.[3]

> *There are too many of us and we all have such entangled lives. We are at the same time constituted of large parts of the people who began us and who tug and pull to continue us, and smaller fragments of the hundreds of others at back-to-*

3. This is the place to indicate that I teach philosophy at an institution called *Redeemer* College (Ancaster, Ontario). The college's name gives me extra reason to reflect on what redemption means; therefore I offer this philosophical essay as an expression of the college's mission. At the same time, I wish to thank the college for a research leave of absence that helped make the writing of this essay possible.

school night, traffic jams, the grocery store, the beauty shop, bus-station rest rooms.

–Shelby Hearon[4]

Is the past my own? In a significant sense it is indeed my own, for I stand at the center of my own recollections. If we consider the past in the limited sense of that which I remember directly, we must conclude that it is surely mine. Even if it includes events of which I was a mere spectator (perhaps a horrified or frightened spectator), my feelings about such events at the time they took place are very much a part of the recollections; indeed, if I had no strong feeling about a certain event at all, it is not likely that I would remember it for long. Most of us would say that we remember only that which involved us personally at a time when we were conscious. (Later I will try to expand the meaning of the term "remember" beyond this strict or narrow sense.)

A bit of reflection reveals that some qualification is needed when the question whether the past exists is asked. Our memories of events in which we participated are also memories about *other people* who were involved in those events with us. Thus, if we affirm that the remembered past is our own, it is not just our own action that we remember. Much human action is *joint* action, and some of it can be called *inter*action. We share the past (i.e. those joint actions and interactions) with others whose lives are entangled and intertwined with our own. Hence we sometimes find ourselves arguing with others about just how things were and what actually happened on a certain occasion. If the past were strictly my own, as my fantasies are my own, such arguments would make no sense.

There is an even more interesting sense in which my past is not just my own. Our memories of the past — even of our personal past — are not independent but need to be supported, confirmed, refined, and buttressed by memory supports[5] of various kinds. Memories do not maintain themselves on the basis of any inherent power: they need to be nourished, and other people help us nourish them. This is one

4. *Hannah's House* (Garden City, N.Y.: Doubleday & Company, 1975), p. 12.
5. See the Glossary for a definition of this term.

of the reasons why many adults who have paid little attention to their brothers and sisters over the years come to value them greatly in old age: those siblings are able to help them nourish precious memories dating back to their very earliest years. To lose a sibling who serves as a memory partner is to lose part of one's own store of memories, part of one's own past. And so we see that we must qualify the thesis that my past is my very own.

> *This miracle that human life is swept clean by the fact that one 'does not remember' is one of the most powerful re-creative ethical forces in existence.*
>
> –Gustaf Wingren[6]

What may I throw away? A lifetime of living leaves us accumulating more goods and possessions than we can sensibly use. We are forced to dispose of things, especially in later years. Sensible people sort through their possessions all throughout life, getting rid of things from time to time. And something of the same process seems to be necessary with regard to experience and our memories of the past. There is so much more that could be retained than we have room for. And so we find ourselves sifting, selecting, throwing away.

The analogy between retained experience (in the form of memories) and possessions is not complete, of course. Throwing things out is more definitive than forgetting. In later years, episodes that we seemed to have forgotten may come back to us, sometimes with remarkable clarity. There is no parallel possibility when it comes to items we have consigned to the trash, however much we may later regret disposing of them. Those items have been burned or buried or perhaps dumped in the ocean.

Focusing, sifting, and attending selectively to what experience holds before us as potential objects for consciousness — all of these are existential necessities, as Nietzsche, in particular, has reminded us in a famous essay.[7] There is a danger of being overwhelmed by the

6. *The Flight from Creation* (Minneapolis: Augsburg, 1971), p. 71.

7. Published in English under the title *On the Advantage and Disadvantage of History for Life*, tr. Peter Preuss (Indianapolis: Hackett, 1980). Also available under

past, to the point that action in the present becomes difficult or perhaps meaningless.

That the process of "cleaning up" our memories is morally and existentially problematic becomes evident especially when it concerns persons. Consider the following.[8]

> ◘ I hadn't seen Barker for twenty years — not since the war. I'd thought of him often, and of how he'd saved my life back there in France. Before that fateful day, I had taken him for an upper-class snob and never knew quite how to relate to him. After he saved my life, I was so frightened by my brush with death and overwhelmed with gratitude that I couldn't think straight — let alone talk. And I never really thanked him — not properly. In a skirmish the very next day Barker was wounded himself, and I saw no more of him. I suppose I could have gotten his address and written to him, but I never got around to it.
>
> And then, one day, I saw him on the street. There was no mistaking him. But you know what? He didn't recognize me at all, and he said he didn't even remember how he saved my life! He didn't deny it, mind you; he just didn't remember it specifically. "So much happened during the war," he told me.
>
> Can you imagine how I felt? Here I'd had this great obligation weighing me down all these years — and now it seems nothing. It's almost got me doubting whether it ever really happened. After Barker brushed me off — he said he was in a rush — I didn't know how I was supposed to feel about it all. But I can tell you I was humiliated. And I wouldn't dare go up to him again.

In this little story Barker has completely forgotten a man whose life he saved during the war. By itself this is not entirely unusual — what is strange is the role Barker played, for a period of some twenty years, in the consciousness of this man. (Already we suspect that the narrator of this story is overdramatizing the events in France twenty

---

the title *The Use and Abuse of History*, revised edition, tr. Adrian Collins (Indianapolis: Bobbs-Merrill, 1957).

8. This essay on narrative will use brief narratives (all of my own devising) to illustrate points. To make the narratives easy to distinguish from the exposition, I have set them in smaller type as indented quotations introduced by the symbol ◘

years earlier: perhaps he is making a largely accidental chain of events into more than it really was.) Did Barker have some obligation to retain a vivid memory of the man he rescued and of the events that day? The same sort of thing can happen with so-called love affairs. What might have been a very intense and keenly remembered relationship for one person may be all but forgotten by the other. ("There were so many girls . . . .")

Do we have moral obligations when it comes to remembering? Sometimes we are hesitant to dispose of certain knick-knacks because we acquired them as gifts. But if the giver is dead and gone, there is no reason *not* to get rid of them, for he will no longer come calling at our home, looking for his gift on the mantlepiece. Perhaps similar social obligations devolve upon us when it comes to memories. "Do you remember the time when . . . ?"

> *O Lord my God, is this not the truth as I remember it? You are the Judge of my conscience, and my heart and my memory lie open before you.*
>
> –Augustine[9]

Is there any truth in the past? We would certainly like to think there is, but the truth of the past does not simply preserve itself. If we are to have access to it and live by it, we have to work to (re)construct and preserve it. The narrative about Barker, who completely forgot about saving a fellow soldier's life during the war, illustrates how easy it is to let the past slip away into oblivion — we can even forget about things that are to our credit.

Embarrassing and shameful episodes are forgotten all the more readily, especially when our fellow rememberers are no longer around to remind us. Or, in many cases, such episodes are gradually altered so that we come to stand in a better light. Stories change in the retelling, just as fish grow larger when fish stories are retold.

We may well feel inclined to condemn people whose memories of shameful episodes in their own lives appear to be none too accurate.

---

9. *Confessions,* tr. R. S. Pine-Coffin (New York: Dorset Press, 1986), Book 5, Section 6, p. 98.

Yet it should be borne in mind that the shame and guilt attaching to something that has happened may be dependent upon interpretation. Perhaps one is ashamed because of the circumstances — or perhaps because the act in question was carried out before the eyes of So-and-so. (Think of a man being humiliated before his own wife.) When such a story changes in memory, it is not so much that facts are denied; rather, what happens is that the circumstances of the episode are reconstructed in a way that places the rememberer in a better light.

An outright denial of the facts is also possible, of course. One way to understand totalitarianism is to view it as the determination to exercise a monopoly on remembering (at least in the social, collective sphere). Those who exercise the monopoly punish deviants who insist on remembering things their own way; they also control the memory supports (e.g. rewriting the history books from time to time) and are able to eliminate such traces of the past[10] as might contradict the official version. Indeed, in its most extreme form, totalitarianism freely invents false memory supports and alters traces of the past. George Orwell has given us a chilling portrait of such totalitarianism in his novel *Nineteen Eighty-Four.*

It is my premise in this essay that there is indeed some truth to be had in the past. (Perhaps we could better say that there is truth *about* the past, for we "have" it in the present.) How is such truth to be attained? My thesis, in brief, is that we should regard our narratives, rooted as they are in memory, in memory supports, in the testimony of others, and in actual traces of the past, as possessing a *prima facie* validity. Those narratives need to be corrected and refined. Sometimes a quasi-Cartesian doctrine of distinctness in recollection seems to hold us back from such a recognition: we overestimate the power and accuracy of personal memory, and at the same time underestimate the value of the memory resources[11] available to us.

I am well aware that the belief that there is truth about the past flies in the face of much fashionable skepticism. In some quarters we often hear it said that claims to historical truth or to an objective

10. See the Glossary.
11. See the Glossary.

knowledge of the past are self-serving efforts at manipulation, or unconscious justifications of some unjust order of things. In this essay I am presupposing that the Biblical imperative regarding truth-telling ("Thou shalt not bear false witness") should be understood as linked with an assurance to the effect that efforts to be fair and honest in talking about the past can indeed be rewarded. It is part of my purpose to spell out some of the factors that contribute to historical truth — and even objectivity.

Yet any commitment to truth in memory and narration requires of us that we become more self-conscious than many of us currently are about the factors and patterns operating in memory processes and in our storytelling. Before we turn to such considerations, a comment is needed about the phenomenon of psychotherapy, and another one about the religious quest to find meaning in the past.

Can psychotherapy help me find the truth? The search for the truth about my own past would seem to have a good deal to do with the twentieth-century pastime known as psychotherapy. The healers we call psychologists and psychiatrists have a number of techniques and tactics at their disposal: some prescribe strong medication, other try to shake up their patients by various psychological — and even physical — shocks, and then there are those who offer what we might call a gentle talking cure. The point of such a cure, presumably, is that a troubled individual, by talking with a sensitive and understanding person, can come to a greater awareness of his own past and can eventually accept it and go on from there. At first glance, then, it would appear that psychotherapy, at least in some of its contemporary forms, is friendly toward our effort to uncover personal truth in the past.

This summary of the situation is a bit too simple, for there are also elements in the world of psychotherapy that have radically undermined the commitment to building integrity by seeking truth about one's past. The suspicion of psychotherapy in our society leaves us with the impression that the revelations which patients make to their psychotherapists contain more than just a few grains of falsehood. To hear some people tell it, psychotherapy is a process of blaming one's woes on others, especially one's parents, who, by the time psychotherapy is begun, may well be dead and are thus unlikely to contradict any slanted revelations their children offer.

The seeming tolerance for fantasy and untruth on the part of many psychotherapists has a great deal to do with the career of Freud. In his early days, when he was first working with the set of theories that came to be known as psychoanalysis, Freud believed that the disturbances of a great many of his patients stemmed from what we now call sexual abuse: his female patients, it appeared, had been forced to engage in unwelcome sexual acts with their fathers while they were still very young. At a certain point, however, Freud came to the realization that many of the stories of childhood seductions by fathers which he had been hearing from his patients were fabrications. Perhaps he should then have abandoned the theories he was developing,[12] but instead he offered a seemingly slight modification: it did not really matter, he decided, whether the reported seductions were actual. What his patients were telling him represented psychological realities for them, whether rooted in memory or in imagination. The long-term result for the Freudian wing of psychoanalysis is that the long process of opening one's inner life to one's analyst lost touch with (historical) reality and actuality: indeed, the patient himself was not even expected to have much sense of what it all meant. He was told that it was best to just say whatever came to mind and let his analyst take the lead in sorting it out.

This move within the world of Freudian psychoanalysis has had a significant impact on our culture. We also see its influence in the popular novels of our time. In both psychoanalysis and the novel, it appears that one is licensed to say whatever pops into one's head, with the notion of a border between truth and fiction openly scorned. In the world of the novel, the omniscient narrator[13] had already been fading for some time (although many readers still feel more comfort-

---

12. Just what Freud did do — and should have done — is the subject of Jeffrey Masson's controversial book *The Assault on Truth: The Suppression of the Seduction Theory* (New York: Farrar, Straus & Giroux, 1984).

13. See the Glossary.

able with his guidance through a story). In many twentieth-century novels, one learns not to accept as fact (within the fictional world of the story) what the limited, biased, prejudiced narrator says: we are to bear it in mind, but we should not be surprised (later in the novel) if an altogether different light is shed on certain circumstances. The narrator is not reliable.

Nourished by the contemporary novel and the rise of Freudian psychoanalysis, then, the garrulous self we encounter today has learned to talk the language of feelings, scarcely making a clear distinction between *events* and how he *feels* about events. Becoming more open and fluid in expressing one's feelings ("getting it out") is widely thought to be a virtue. And it may indeed be good for the soul to talk freely about one's past, especially those parts of the past that give rise to continuing guilt and anxiety, but there is also an older mode in which such talk can be couched — the mode of confession. To quite an extent, psychoanalysis and the novel have taught us a new way to talk about ourselves. What we must ask is whether such discourse is better for us than the traditional Christian practice of confession. "Search me, O God, and know my heart! Try me and know my thoughts! And see if there be any wicked way in me . . ." (Ps. 139:23–4).

> *The Greeks . . . thought that even the gods could not change the past; but Christ did change it, rewrote it, and in a new way fulfilled it.*
> –Frank Kermode[14]

How can the past be redeemed? My thesis is that the past is redeemed in words (storytelling, or narration) as well as in actions. In some situations, we succeed in "making something" of the past through both words and actions. Consider the following example. (The third-person narrator is an older brother of "Jill.")

■ For about the first six months that Jill was going out with Jack, we didn't think all that much about their relationship. She was only

14. *The Sense of an Ending: Studies in the Theory of Fiction* (London: Oxford University Press, 1967), p. 47.

eighteen, and it didn't seem all that serious. But then one day it all came to a head: Jill announced she was pregnant, and Jack had disappeared without a word! Not even Jill knew where he'd gone, and she didn't get as much as a postcard from him.

Well, I didn't expect we'd see much of Jack after that, but two months later he showed up again and volunteered to marry Jill. I was against it: she would have been better off putting the baby up for adoption and continuing her education. She'd made a mistake, but she could still have gotten back on track. Instead she married him (although she didn't look all that eager to me), and they even made something of a go of their life together.

Their own children are teenagers now, and naturally they ask difficult questions every now and then. I was at their place for supper one day when I heard the current version — how Jack and Jill were so deeply in love that they got "carried away." It was wrong, they admitted. But they didn't say anything about Jack's two-month disappearance. And I didn't have the heart to bring it up.

This narrative represents a slightly cynical interpretation of how two people sought to redeem their own past actions by word and deed. If Jack were telling the story (let's suppose he's addressing his own teenage children), it would have a different flavor.

Your mother and I were quite young — too young — when we got seriously involved. Then one day she told me she was pregnant. I panicked and went away for a while: I just couldn't face up to my own responsibilities toward her. I needed time to think. But what kept staring me in the face while I was away from her was how much I loved her. I wanted to spend my life with her, although I didn't really want our marriage to begin with a pregnancy. But that couldn't be helped any more.

Anyway, I went back and told her I couldn't live without her and asked her to become my wife. I've never been sorry.

In this version of the story Jack's disappearance is not denied, but we are certainly given a much more sympathetic and benign picture of it. The point that needs to be pondered is not which of the two narratives is true (or perhaps more true) but what limits there are to the enterprise of redeeming the past, for this is indeed the project in which Jack is engaged. On the one hand, there are cynics who laugh at such stories as Jack's and dismiss them as retrospective falsifications of what actually happened. On the other hand, many people would sympathize with Jack's response to his predicament. (Jill also has a story, of course, but I do not propose to introduce it here.)

If we find Jack's approach legitimate, what are the constraints that should be placed on his later (retrospective) narrative reconstruction of what happened? To what extent does his eventual success in the marriage justify a neat and tidy portrayal? Surely other possibilities must have swirled through his mind during his two-month period away from his pregnant girlfriend: do they also have to be reported? Or is the winning "strategy" the only one that needs to be mentioned?

The strong element of retrospection in this story is reminiscent of Christian theology. A Christian (especially Reformed) understanding of the Bible usually stresses that earlier events are to be understood and narrated in the light of later events (their fulfillment). One does not preach about the story of Abraham's willingness to sacrifice his son Isaac without mentioning Christ, in whose case the sacrifice of the beloved son was actually carried through. Even the sordid story of Judah and Tamar (Genesis 38) is said to be about Christ in some sense. The "felix culpa"[15] strain that runs through Christian thinking is of a piece with Jack's effort to redeem the past by subsequent action and by a later telling of the whole story in which the outcome largely determines the story's shape. If God can redeem the past, shouldn't we try to do the same to the limited extent that we are able?

15. The "happy fault" is man's fall into sin, which made the glorious plan of redemption necessary.

Chapter 2

Reduction and Simplification

Memory is the opposite of thought. . . . the thinking man keeps his mind open while the remembering fellow narrows and simplifies until the memory becomes as crude and bald as a Hallowe'en lantern.

–Julia O'Faolain[1]

How much can we remember? The notion that memory is a faithful record of all that we have experienced is inadequate, as we soon realize when we think carefully about ordinary human limitations. Just as there is only so much room — and no more — in a computer data-storage device, there are limits on how much information we are able to retain in what we call memory.[2] One of the reasons we forget things (e.g. faces once familiar to us) is that a host of subsequent experiences and impressions appear to have crowded out some of the earlier material.

Since memory is in danger of being overwhelmed by a flood of new material, it must make use of certain devices and strategies to organize and arrange what we do manage to retain. There are quite a number of these devices and strategies, and some are probably not well understood. But one of which we are fairly well aware is a process of reduction. Often we find that we have, in some sense, remembered an experience or a set of connected events as a whole, even though various details and episodes within the set seem to have

1. *No Country for Young Men* (New York: Carroll & Graf, 1980), p. 65.

2. Although the term "memory" is widely used, it does require some justification, which I will provide in the next section.

slipped away. A vacation journey you took fifteen years ago may seem fairly clear in its general outline: perhaps you could still trace the route on a map. But surely there would be little episodes within that journey and vacation which you had forgotten. Perhaps hearing them retold by someone who was with you would bring them back to the "forefront"[3] of memory. Yet there might be others of which you would say, "Did that really happen? I have absolutely no memory of that." Within a marital dispute, one partner will sometimes deny having said any such thing as is now thrown up in his or her face, while the other partner seems to remember it vividly.

Sometimes a distinction is made between short-term and long-term memory. Such a distinction is at least a promising beginning in solving the theoretical problem we face here. It does appear that as we experience things, we retain them in a reasonably full way for a few seconds and then make some sort of unconscious decision as to whether to retain them on a longer-term basis. Sometimes we are listening casually to a conversation or perhaps to a radio when suddenly we are jolted within and wonder, "What was that I just heard? Did I hear a reference to my old friend So-and-so?" At that point we seem able to "play it back"[4] to determine whether we really heard what we thought we did.

It appears that further decisions about *retaining* items as opposed to letting them slip away are made later on. (The process by which items of past experience become completely inaccessible is not as decisive as setting out the trash early in the morning and noting later in the day that it is gone.) Perhaps an adequate model would have to include *three* stages or levels of retention. The difference between the second and the third is not well understood, although it does seem fairly certain that any event to which strong emotions are attached stands a good chance of being remembered for a long time. But even

---

3. This is not an exact or theoretically justified term. But it does seem that we have ready access to some items in memory, while others emerge and stand on their own legs as memories only after we have been prompted by someone. We will see later that we have reason to be suspicious of such "prompted" memories.

4. Another theoretically unjustified notion. Many people compare this ability to the workings of a tape recorder.

this generalization could be called into question by the findings of psychoanalysis: it also appears that we *repress* certain frightening experiences which continue to affect us but cannot be called forth on the same basis as ordinary memories.

The psychoanalytic input does seem to suggest, however, that there is a practical motive behind much memory organization. The person who represses memories cannot cope or function properly without doing so: we could speak here of an *existential* necessity. Ordinary memory failings (e.g. difficulty in matching a familiar face from the past with a name) are frequently tied to practical needs and situations as well: there is a *context* to acts of memory retrieval. If I attend a reunion of high school classmates, I am mentally geared to certain faces and draw on a certain "portion" of memory tied to my high school years. But if an old friend from high school pops up where I would least expect him, I probably will not recognize him. When I return to my teaching duties after a summer away from the students, I "refresh" my memory by looking through the pictorial directory of students published the year before: many of my students have "slipped away" from me over the summer. What we seem to keep in the "forefront" of memory (that to which we have ready access) is mainly that which we are likely to need and use.[5]

Such considerations help us see why homunculus[6] theories of mental life have enjoyed a certain popularity. It sometimes seems easiest to explain the working of memory by reference to a remarkably practical "little man" who keeps things neatly organized in the "warehouse" of our memory. He sorts, throws things out, and compacts what is left for easier storage. But he's only human and does not

5. Even the retrieval of words for the sake of making English sentences seems to follow this pattern. Sometimes I find myself wondering why I used a certain word in conversation, only to realize that I had used the same word (perhaps with an entirely different meaning) in the previous sentence. The word was ready to hand, so to speak. Or the word may just have been used by my conversation partner.

6. A homunculus is a little man. A homunculus theory tries to explain agency by positing an agent (i.e. a little man) within the agent, just as a small child might suppose (as one of my nephews did) that there are little people inside a radio producing the voices and music.

always remember where he has put things.

Selection in everyday experience. The problem of abundance in our mental life does not *begin* at the level of memory: we already confront it in direct or primary experience. Under almost all circumstances, we find ourselves in the experiential presence of much more color and sound and shape and aroma than we are able to attend to. And as we make decisions to zero in on one item or feature or quality at the expense of others, we often change our physical positioning. We approach an object to look at it carefully, or we cock our head to one side to be able to tune in on a conversation going on behind us at a stand-up reception.

When we consider the sense of touch, we find that it is more localized than vision or hearing or the sense of smell. We usually use our fingers and hands to get the feel of things, although other parts of our bodies are also capable of great sensitivity. (Sexual experience draws on those sensitivities.) Touch, then, is even more a matter of choice than is hearing or smell. Yet a degree of focus can be attained in both of those sensory systems as well: we listen carefully to one strand in a piece of music, or try to isolate one aroma in an appetizing dish that is placed before us. And in vision, of course, focus comes fully into its own: we can look at one item so intently that we become oblivious of other items in our visual field. "I have eyes only for you," the lover says to his beloved.

If primary experience is so selective, surely memory, which is first of all a retention of the content of that experience, is shaped by the same factors of selectivity. The limitations of experience (in the face of cognitive abundance) are also constitutive for memory. And this leads me to ask just what the boundary is between experience and memory. John Locke and David Hume, the classic empiricist philosophers, had distinct views on this matter, but the line they drew is a little too neat to be accepted today. Their conception of experience was too passive: it was as though sense data[7] simply imposed them-

7. Here I am using a later term. Locke spoke of ideas, and Hume of impressions, as the primary givens in experience.

selves on our cognitive apparatus. The constructive role of the human mind was not given sufficient attention in their philosophy. But even if, for the sake of the current exposition, we do not challenge the empiricist position regarding the relationship of memories to the original impressions making up experience, the selectivity and focus that is inevitably a part of experience must be recognized as built into memory from the outset. The conclusion, presumably, is that we should regard neither experience itself nor memory as some sort of transcript or photographic reproduction of elements of reality.

Why speak of memory at all as something separate from experience? And why distinguish memory from imagination? Could it be that these terms are dubious reifications[8] that need to be challenged and cast aside? Don't memory and imagination intertwine to the point that it would be unwise to disentangle them conceptually?

To sort out these questions, it may be helpful to consider I. M. L. Hunter's definition: "Memory, in its most comprehensive sense, refers to the effects of a person's past on his present."[9] On this view, memory is the operative retained past. To defend such a conception of memory is really to offer an answer to the question of the reality of the past. Is the past a factor in our life? Yes indeed, for it is present to us in the form of memory. To allow all of memory to be placed under the heading of experience is to deny the past — there is only the present.[10] But this conception does not demand that there be

---

8. A reification (thing-ification) is a theoretical construct or concept that may be useful for discussion but does not directly correspond to any separate item or given in human experience. Common sense tends to posit its separate existence; that is to say, it is inclined to regard a reification spoken of as a concrete existent. Some examples: the gross national product, the German mind, the Renaissance, artificial intelligence, the patriarchy.

9. *Memory,* revised edition (Baltimore: Penguin Books, 1964), p. 19.

10. This has indeed been an option in philosophy of history: all historical knowledge is then present knowledge. R. G. Collingwood writes: ". . . all historical thought is the historical interpretation of the present . . ." ("The Limits of Historical Knowledge," in his book *Essays in the Philosophy of History,* ed. William Debbins [New York: McGraw-Hill, 1966], p. 102) and maintains that the historian's goal is "knowledge of the present" (see "The Philosophy of History," in *Essays,* p. 139). Collingwood interprets the early Benedetto Croce as teaching that all history is

demonstrable qualitative and "phenomenological" differences between memory and experience. (Hume thought that original experiences were more vivid than memories.)[11]

It will also prove necessary to distinguish memory from imagination (see Chapter 7 below). For the moment, suffice it to say that imagination is personal and individual in a way that memory cannot be, however intimate and embarrassing some of our memories may be. It is a characteristic of memory that it embodies a reference to the real world of space and time which I inhabit with many other human beings, who thus enter into my memories as real people. The imagined world (even though it may resemble the real one, as the imagined world of many a novel surely does) is *not* the real world. An imagined character or incident may *seem* real (we sometimes use the term "realistic" here), but it is part of a *private* world. Even our memories of past fantasies (which may bother us and come up for discussion in psychotherapy) qualify as memories only because of their link with the real world: they are part of the inner life of a real person and affect his outward behavior.

Things aren't the way they used to be — if they ever were.
–Bobbie Ann Mason[12]

Texture and vividness. For some thinkers the problem of memory and its accuracy appears to be largely a question of restoring the vividness and intensity of original experience. It is then assumed that the memories that are vivid, richly textured and detailed, the ones in which we seem to float as though outside of time, are the ones most worth cultivating; indeed, some regard them as a recovery of exactly "how things used to be." Such a conception of memory is often

---

"contemporary history" (see *The Idea of History* [New York: Oxford University Press, 1956], p. 202).

11. See his *Treatise of Human Nature,* Book 1, Part 1, Section 3, where he deals with the relation of memories (which fall under the heading of ideas in his terminology) to original experiences (impressions) and the ideas of the imagination.

12. *Spence and Lila* (New York: Harper & Row, 1988), p. 108.

associated with Marcel Proust.[13]

In terms of their intensity, such memories are like snapshots. And their value is like that of photographs generally: they do provide a sense of how things were, but they are episodic at best and do not give us a sense of coherence among the past events which we remember. For Proust, memory cannot go beyond such a sense or impression. Georges Poulet writes: "Perhaps the greatest difficulty of the Proustian enterprise consists in the fact that for him knowledge can never cease to remain impression. It is possible to know only that which can once more become immediately contemporaneous to the heart. Now, 'knowledge in these matters being intermittent and incapable of surviving the effective presence of feeling,' the result is that for Proustian thought the *knowing* as well as the *being* finds itself bound to a world essentially ephemeral and intermittent . . . ."[14]

The kind of memory I am concerned with in this essay has a great deal to do with our sense of ourselves, of who we are. And that sense of identity is bound up with a remembered story or history. In linking the terms "story" and "history," I am not suggesting that they are synonymous; what needs to be noted is that they overlap. History is normally understood as a narrative record of important events that actually happened, whereas a story a person tells about himself can be either true or made-up (fiction). When an ordinary person tells the (true) "story" of his life (as opposed to speaking of "history," as a statesman might do when writing his memoirs), he is acknowledging that he is not widely known or thought to be "historically important." Whether a given individual is being too modest in this regard is a matter about which historically informed people might disagree. Part of my thesis in this essay is that we should not be overly modest when it comes to thinking of ourselves as part of "history."

Now, a story simply *must* contain a great deal of coherence and

---

13. In *The Remembrance of Things Past,* Proust writes about incidents of vivid recall that are set off by physical stimuli, the first of which is the celebrated "madeleine" cake: see Vol. I, Moncrieff translation (New York: Random House, 1934), pp. 34–6.

14. "Proust and Human Time," in *Proust: A Collection of Critical Essays,* ed. René Girard (Englewood Cliffs, N.J.: Prentice-Hall, 1962), p. 170.

continuity if it is to be a story at all. (To what extent such coherence is remembered — as opposed to being constructed after the fact — is in good measure the question I wish to probe.) Proustian recall falls outside this framework. We draw on it (that is to say, those who possess this gift, which is not evenly distributed throughout the human population) to supplement other memory sources and aids, but it does not by itself represent the heart of memory, or even the key to memory. Perhaps it is overemphasized in our culture because of our preoccupation with the contemporary novel, which so often utilizes keenly remembered scenes from childhood.

It should also be noted that Proustian recall, however pleasurable it may be in itself, does not contribute much to life on an everyday practical level. If memory is to be justified on practical grounds as a set of operations that helps us manage day by day, Proustian recall hardly qualifies, for it often immobilizes those who bask in it, as we can see from Proust's own novels. It tends to come to us somewhat later in life, when we are usually less preoccupied with the future and with getting ahead. Proustian recall does confer a certain aesthetic delight and satisfaction (for it is surely pleasant to re-insert oneself into the world of childhood), but it is not to be sought for the good it does us.

Perhaps the conclusion to be drawn is that there is a contrast and trade-off between texture and vividness, on the one hand, and coherence, on the other. Coherent memories, like coherent perceptions, will tend to include less in the way of texture and raw feel: they will be more rounded off, and therefore easier to assess when brought into consciousness again after lying unused for some time in the "storehouse" of memory. A powerful experience can overwhelm and immobilize us, and so can Proustian recall. But memory in Hunter's sense (i.e. the effect of our past on our present) usually comes equipped with packaged lessons. Using the term "moral" in a broad sense, we could say that there is a strong moral and practical element in experience. That's why we often ask, after hearing someone relate an interesting episode from his own life, "So what's the moral of the story?"

> *. . . there are no certain indications by which we may clearly distinguish wakefulness from sleep . . . .*
>
> –Descartes[15]

What about dreams? The suggestion that there may be a trade-off in our experience between vividness and coherence gets a bit more support when we turn to dreams. Some people maintain that the vividness of at least some of our dreams requires us to admit that they are indistinguishable from waking experience. But since one cannot enter another person's dream to ask him whether he is dreaming or awake, what this suggestion turns into by way of a possible experiment is to ask someone, when he is awake, whether what he is experiencing at that moment could be a dream or a nightmare.[16]

Descartes' famous claim that the dream state cannot be distinguished from the waking state when it is examined by itself needs to be challenged. I do recall a couple of occasions when I distinctly concluded, during a dream, that I was only dreaming and should therefore not concern myself overly about what was happening. On one such occasion I was late for school. In this dream, which occurred when I was in grade five, I was worried about being punished for being late, as I had been on other occasions. I ran in order not to be any later than necessary. But as I was running I realized it was only a dream, and so I slowed to a walk. Sometimes I find myself in a pleasurable dream-like state just when the alarm clock rings. If the alarm is quickly turned off, it is sometimes possible to re-enter the dream state and milk a bit more pleasure from it.

We also form memories of our dream experiences, but we generally do not retain them for long. If our dream involved the death of a loved one, there may well be an emotional carryover throughout the day: we feel sad, even though we know there is nothing to be sad

---

15. *Meditations,* in *The Philosophical Works of Descartes,* Vol. I, tr. Elizabeth S. Haldane and G. R. T. Ross (Cambridge: Cambridge University Press, 1967), p. 146.

16. When this question is asked by a professor delivering a boring lecture in his Introduction to Philosophy class, it usually sparks some interest: maybe it really is a nightmare!

about. The feeling of sadness we then experience is akin to what we might feel from beholding a powerful artistic rendering of the death of someone's loved one. But often we cannot remember a day later what it was that we dreamed.

The reason we usually forget such dreams before long is practical: we distinguish dreams (and the memories we form of them) from real experiences (and their corresponding memories) on the basis of *coherence*. Normally we do not link up the dream experiences with real ones: they do not form part of the flow of life, and so we let them slip away as not relevant. We would be able to retain them longer if we judged them relevant or important. I remembered the dream about being late for school because it taught me something about the nature of dreams.

Yet there is psychoanalysis to consider here. Both Freud and Jung, in their respective ways, maintained that dreams serve to tell us certain things about the fears, aspirations and desires of people; in short, dreams have a *meaning* and need to be interpreted. But such an approach to dreams moves them into the area we usually call imagination. To accept such a view is to conclude that dreams are not primary experiences or memories of what has happened but *responses* to the events and dynamics in our life. They bear messages that are important for how we live our lives. And their very vividness testifies against any suggestion that we should treat them as memories, for most memories lose their "color."[17] Yet there are some dreams that confuse us on this score because they draw on real experiences (such as frightening episodes in the past) and thus seem to us a mixture of memory and imagination.

All acting requires forgetting, as not only light but also darkness is required for life by all organisms.
–Nietzsche

17. It is significant that Hume emphasized that the products of imagination are much more vivid than our memories: see his *Treatise of Human Nature,* Book 1, Part 1, Section 3.

God is merciful in what He sometimes lets us forget.
–Chaim Potok[18]

Nietzsche's lesson. Most human beings who are no longer young are familiar with the failings and shortcomings of memory. For this problem, as for so many others, various remedies are advertised, e.g. programs for increasing one's "memory power." The implication seems to be: the more memory, the better. We do not find people promoting ways to *decrease* memory. Yet we do face the necessity of decreasing our memory store, of filtering out of our ordinary experience many of the items and aspects that do not pertain to our needs and interests. Thus, if there is an imperative to *remember,* there also appears to be a need to *forget.*

This point has been made most eloquently by Nietzsche in the essay referred to in Chapter 1. Writing late in the nineteenth century, Nietzsche was opposing the Hegelian pretension to a complete and ever growing consciousness of the past (however compressed and *aufgehoben*[19]). The "historical universalist," as Nietzsche calls him, finds traces of himself everywhere in human evolution and history as, "high and proud," he looks back from his perch on the "pyramid of the world process" and cries out: ". . . we are the goal, we are the completion of nature." Nietzsche's response: "Overproud European of the nineteenth century, you are mad!"

Nietzsche's analysis underscores the moral-practical conception of memory I have been developing, as we see especially from the title of his famous essay on the burdens of memory and the historical sense: "On the Advantage and Disadvantage of History *for Life*" (italics mine –TP). The needs of *life* must be uppermost in our consideration of the role of memory, of the historical sense. Nietzsche discerned in his own time, with its excess of "historical education," a kind of "inborn grayheadedness." The result of this condition is that we become suited

---

18. *The Gift of Asher Lev* (New York: Alfred A. Knopf, 1990), p. 58.

19. *Aufhebung* is Hegel's term for the filtering and elevation to a higher level which human experience and culture (including truth claims) undergo in the course of the historical process.

only to the activities of old men, which are "to look back, to tally and close our accounts, to seek consolation in the past through memories." Truly great action in the present, action in the service of *life,* requires that we *forget* resolutely: "As the man of action, according to Goethe's phrase, is always without conscience, so he is also without knowledge; he forgets a great deal to do one thing . . . ." Yet forgetting is not as easy it at might appear. In praising "the unhistorical," then, Nietzsche is trying to help us summon "the strength of being able to *forget.*"

To avoid drawing one-sided conclusions from Nietzsche's lesson, we must look once more at the title. Although his essay is widely read as an attack on any preoccupation with history and the claims of the past, his title does leave room for benefits to life which flow from memory and the historical sense. Man distinguishes himself from the animals in his capacity to bear a relation to the past, from which he must continually select that which he can use in the present. (The cow forgets, and thus is always content, "enthralled by the moment.") Even the ancient Greeks, to whom Nietzsche points as setting the proper example, became "enrichers and increasers of the inherited treasure."[20]

> *I own many books full of my annotations, proving that I once read them, though I have no memory of it.*
> –John Updike[21]

Memory supports. The discussion of memory is often hampered by what might be called a quasi-Cartesian conception according to which our memories can provide *internal* evidence of their authenticity. Either we remember something (clearly and distinctly), or we don't. When we mutter, "I have a vague recollection of some such incident," we are either being polite or fooling ourselves. This conception, I maintain, needs to be set aside so that we can recognize the massive role of memory supports in everyday life. The way we nor-

---

20. The quotations from Nietzsche are drawn from pages 10, 50, 44, 11–12, 8, 62, and 64 respectively of *On the Advantage and Disadvantage of History for Life* (Chapter 1, Note 5).

21. *Self-Consciousness: Memoirs* (New York: Alfred A. Knopf, 1989), p. 247.

mally think of memory does not fit in with this quasi-Cartesian conception at all.

Suppose someone walks up to me after I fail to address him by name and asks, "Don't you remember me?" What does this question really mean? In the strictest sense we could take the question as a challenge: the person is probably a former student of mine, and what I must now do is match his vaguely familiar face with a name. Sometimes I am able to do so immediately, but in many cases the name does not come to me. After a summer away from the students, a name sometimes eludes me for days, even though I can recall many things about the student who wants to know whether I remember him.

Another possible interpretation of the question I am being asked is this: do I have ready access to information about the person before me, information that has to do with my relationship with him when he was my student? Surely it is my obligation to be able to come up with information about him, since he may be asking for a recommendation for further studies or employment. To remember him means, in part, to be able to find him quickly in my records, look up his grades in my courses, and perhaps trigger the recollection of incidents that involved him and are relevant to his request for a recommendation. An old college yearbook might help in this regard. Now, suppose I look him up and trigger some recollections — does this count as remembering him? My answer is yes.

Not long ago, a former student whom I had not seen for years telephoned me long-distance while I was working at my computer at home. He could not put the facial recognition challenge to me, of course, but he told me his name and seemed politely curious whether I would manifest some sense of who he was. (I had taught him in another institution in another country.) While exchanging pleasantries with him, I quickly opened a computer file that lists former students in alphabetical order, and I soon knew which semester it was that he had studied with me, which course he had taken, and so forth. He seemed impressed. I did not let on that I had looked him up.

I maintain that I had remembered him, in an ordinary sense of "remember." Nor is this conception of what it means to remember dependent on computers. The most a computer does is to speed up access to records. Before I became a computer user, I "remembered" in the same way by making substantial use of written records. If a student asked for an appointment to see me, I might consult my records on him before he arrived.

The point I am making about memory is often formulated on a more general level in connection with the concept of mind. The (anti-Cartesian) thesis is that what we call mind is not restricted to human consciousness; rather, the rational and intelligent organization of our way of life embodies mind (or intelligence) in many of its aspects. Culture can be regarded as an extension of human intelligence, intentionality and consciousness.

In similar fashion I would argue that memory must be understood as including our records and other memory supports. I normally do not go to the supermarket without a shopping list in hand. When my wife asks me whether I remembered to buy this or that item she requested, I answer yes. Surely no one would argue that I am entitled to answer yes *only* if I carried the list in my head and did not commit it to paper: the fact that I brought home the items requested ought to suffice. A colleague may ask whether I have ever read book such-and-such. I reply that I have not. Later I discover a copy of it on my shelf, with my name written in it and marginal notations in my own handwriting. Am I entitled to go back to that colleague and say, "After I returned to my office, I remembered that I had once read that book after all"?

Those who are inclined to answer the latter question in the negative would do so on the basis of a certain idealism about memory, an idealism that lives especially among the young. When I remember former students by relying on records, I normally do not reveal this to them: they seem happier if I appear to remember them by drawing on consciousness alone. Likewise, a man who remembers his wife's birthday by relying on written records may find himself facing disapproving frowns; at the very least, he does not get to call himself thoughtful. For similar reasons, a person may be put out if you make an appointment to spend time with him and inscribe his name on an appointment calendar. Yet older people are usually *not* offended when you say, "What you've just told me is very interesting and important —

I'm going to write it down so that I don't forget it."

My position, then, is that the problem of memory should be integrated with the question of record-keeping. How can records best be organized and maintained? How can one maximize quick access to them? The person who excels in this area should be said to be good at remembering things and people. I keep track of the novels I read. Now, I am not good at remembering names, but if someone asks me for the name of a novel I once talked about in class and gives me some indication as to what I said about it, I can usually find the title and author quite quickly in a computer file.

There is an additional difficulty with this understanding of how memory actually works. For me it is not really a difficulty, but some people would resist my view because of it. The problem is this: reliance on written records as part of memory allows material that has been provided and guaranteed by *others* to be part of *my* memory. An example from everyday life can illustrate the issue nicely.

We need telephone numbers to get into contact with people. We may carry some numbers around in our heads, but most of us rely on telephone directories to some extent. Of course we may have a personal list or sheet of phone numbers, but we do make use of directories put out by the telephone company, the office at work, our local church, and so on. The question "Do you remember So-and-so's phone number?" is in effect the question whether you can supply the number quickly and without much effort. If you say, "Yes, I have it right here," and then quickly look it up on a sheet you keep on your desk, have you *remembered* the number? I see no reason to insist that this should not count as remembering. I sometimes recite a number from memory when someone asks, and then quickly check it in a directory to ensure that I haven't made a mistake.

Whether my memories are mine alone (as opposed to being partially derived from others) is a significant question. I will have more to say about it in the next chapter, when we discuss mediated knowledge, experience and memory.

Memory is the great organizer of consciousness.
It simplifies and composes our perceptions into

units of personal knowledge.

–Suzanne K. Langer[22]

Mnemonic formulas and orality.[23] To appreciate just how fully information storage devices have become part of our culture — indeed, of our mental life — one need only compare our situation with that of people who live in an oral society. Memory in the narrower sense of what people can recall by "racking their brains," so to speak, is the only real repository of knowledge within an oral culture. To keep alive what knowledge there is, a great deal of recitation and repetition is needed. The differences between oral man and literate man[24] which result from this need to repeat have been ably explicated by Walter Ong,[25] Eric Havelock[26] and others.

The key to memory in an oral society is pattern. Just as telephone numbers have a certain rhythm (one learns to put them into spoken words in a way designed to maximize their memorability), many items of knowledge which one might wish to retain (e.g. the family tree) can be expressed in poetic tables and patterns. Ong calls such patterns "mnemonic formulas."

It is the pattern — and not just the items of knowledge which it helps to transmit — that must be passed from generation to generation. Ong observes: "Oral cultures necessarily place a premium on standardization or fixity, for in the absence of writing the major noetic effort of a society must be not to seek new knowledge but to retain what is known, not to let its scant, hard-won store disintegrate or slip into oblivion." What was known was always in danger of falling away. Ong writes: "Any culture knows only what it can recall. An oral culture, by and large, could recall only what was held in mnemonically serviceable formulas. In formulas thought lived and moved and had its being." The overall tendency here can be stated in simple terms: "The

22. *Feeling and Form* (London: Routledge and Kegan Paul, 1953), p. 263.

23. See the Glossary entries for both these terms.

24. See the Glossary on literacy as contrasted with orality.

25. See especially *Orality and Literacy: The Technologizing of the Word* (London and New York: Methuen, 1982).

26. See *The Muse Learns to Write: Reflections on Orality and Literacy from Antiquity to the Present* (New Haven and London: Yale University Press, 1986).

human mind always likes to simplify."[27]

The result is that retained knowledge in an oral society is both simpler (nuances and shades of meaning are less developed) and more personal. A person's knowledge is then very much his own. Elements of this oral emphasis are still to be found in religious education, where recitation and question-and-answer formulas intended for memorization are widely used. Indeed, the notion of "confession" (that which one affirms and says along with others) hardly makes sense apart from (residual[28]) oral emphases.[29]

In a society dominated by writing, we stand in a somewhat ambivalent relationship to orality. It is widely realized that the oral transmission of information is inefficient: it would be quicker to put the messages down on paper and hand them out (or perhaps distribute them through a computerized system of electronic communication). But the question that needs to be asked is: what do people do with messages and memos that they receive on paper? Very often they file them away in case they are needed later, sometimes without as much as a cursory glance below the heading. When people in organizations express a preference for information on paper, it is often with the semi-conscious intent of not reading it.

And here we bump up against the major problem with literate memory (as opposed to old-fashioned oral memory, with its prodigious

---

27. *Rhetoric, Romance and Technology: Studies in the Interaction of Expression and Culture* (Ithaca: Cornell University Press, 1971), pp. 261, 274–5, 187.

28. Ong and others maintain that there are long-term effects of the oral mind that continue to manifest themselves in a society long after writing has come into widespread use. Those effects are referred to as "residual orality." See *Orality and Literacy,* p. 26; see also *The Presence of the Word: Some Prolegomena for Cultural and Religious History* (Minneapolis: University of Minnesota Press, 1981, first published in 1967), pp. 22, 54, 220, 227, 233–4, 257, and "Typographic Rhapsody," in *Interfaces of the Word: Studies in the Evolution of Consciousness and Culture* (Ithaca and London: Cornell University Press, 1977), p. 151.

29. An interesting theme in the writings of Ong is what impact the rise of literacy and the corresponding decline of orality will have on Christianity: on this issue, see especially *The Presence of the Word,* pp. 179ff, 316ff. This issue helps us understand certain differences between Roman Catholics and Protestants: see pp. 264ff, 272–3, 284–6. Ong himself is a Roman Catholic.

feats of recitation). On paper, in books, and in computerized databases, we possess or have access to incredible amounts of information, but we don't know how to use all that information effectively. A village elder in an oral society may be ignorant by our standards, but at least he has a fairly firm grasp on what he does know. By contrast, a man in our society may have the entire *Encyclopaedia Britannica* sitting on a shelf in his living room (as I do) and still be essentially ignorant. If the richest man in the world bought the entire Library of Congress with the intention of using it as his personal library, he would still face the question how all that information could most effectively be assembled and prepared for his perusal and absorption. Likewise, the President of the United States has enormous intelligence-gathering and research facilities at his disposal: he is never lacking in information about the countries he must deal with. But the massive reports that circulate in government offices do not easily translate into knowledge in the President's head: he needs a staff of aides to read those reports and to brief him in turn on what they say. Once we understand what a problem the massive amount of information available to us can pose, we see why there is a continuing place for the small general library in our society.

The realization with which we are left after this consideration of orality is that we remember in patterns and formulas, just as we tend to *see* the world in certain patterns. There is not a strict line between experience itself and memory when it comes to the patterns that pervade cognition. Something very unusual, something utterly anomalous, cries out for interpretation and regularization. If it is a phenomenon that quickly passes away, we think hard about it until we decide that it was an instance of such-and-such. We do not seem able to hold it in consciousness for long as utterly unique. And once it is interpreted, some of its detail and texture and vividness begin to fade away. When Hume concluded that memories were somewhat faded in comparison with original experiences, it was in part because our memories have undergone this process of reduction and regularization.

The more words Abel learned, the more things had meaning; the

more meaning, the more memory.
–Joanne Greenberg[30]

Things that elude our verbal grasp are hard to pin down and almost impossible to remember.
–Diane Ackerman[31]

Semanticized memory.[32] The conclusion toward which we have been moving is that memory — even more than experience — is thoroughly semanticized, that is, pervaded with interpretation, with meaning that serves to render retained experience more memorable. This feature of memory shows how much it is bound up with our basic intelligence: we try hard to *understand* what we remember, and sometimes we overinterpret. The danger to which we must be alert as our memories are being "edited"[33] is simplification.

Memories are not inert. Therefore the process of semantic regularization does not end when an experience is behind us: we continue working with the retained contents of our consciousness to render them more still sensible. Sometimes a tape recording that is supposed to give us some knowledge of an event turns out to be a disappointment because we cannot make much sense of the sounds it gives us. Although such a tape can be enhanced in terms of its comprehensibility, the recording device as such does nothing to render its contents meaningful.

Memory, with its drive toward meaning, is different in this regard, for it is inherently unstable and subject to change (even falsification) — especially when we are contradicted by others in our account of what actually happened. Eyewitness testimony is subject to retroactive influence and is not quite so reliable as many people in the criminal justice system suppose.[34] In our memory of what we have *seen*, we can be influenced by suggestions that are deliberately intended to under-

---

30. *In This Sign* (New York: Avon, 1972), p. 24.
31. *A Natural History of the Senses* (New York: Random House, 1990), p. 289.
32. See the Glossary.
33. See the Glossary on "edited" memories.
34. See Elizabeth Loftus, *Eyewitness Testimony* (Cambridge: Harvard University Press, 1979).

mine and distort what we remember.[35]

If there is a conclusion to be drawn, it is this: we should not be too stubborn in insisting that a perusal of our own consciousness reveals that such-and-such happened on a given occasion, and that the circumstances were thus and so. Because we "edit" our memories and can even be tricked in such a way that we wind up falsifying them, we may need to be corrected by the recollections of others and by physical evidence (traces of the past) relevant to the event we are trying to reconstruct. And there are still other distorting factors to be considered, as we will see in the chapters dealing with narrative. The strength of memory (i.e. its drive to semanticize experience as it prepares it for retention) can also be its downfall.

---

35. See I. M. L. Hunter, *Memory,* especially pp. 259ff.

Chapter 3

Mediated Experience

> *. . . and then there was a vestry meeting afterwards at the rectory. I remember the vestry meeting. Or at least I think I do. I remember it almost too well. The trouble is, I am the kind of person who can picture scenes created by other people's words.*
>
> –Gail Godwin[1]

Primary and secondary memory? Common sense tends to assume that a distinction is needed between two senses of the word "memory," which we might call primary and secondary memory. I do not claim that these terms are widely used: if people were asked about the distinction I am pointing to, they might instead speak of memory in the *strict* or *strong* sense as opposed to *derived* (secondary) memory. The basis of this distinction is then a corresponding distinction within experience itself.

Consider the following example. During a family discussion of funny things that happened long ago, one of the children is asked whether he remembers the episode that was just mentioned. He says he does. But then his sister objects: "You couldn't possibly remember that. You weren't even two when it happened! You only *think* you remember it because you've heard the story so many times. I'll bet you were in bed when it happened." (In some cases children claim recollections of events that took place before they were born.) The child's memory of the episode in question is not genuine, then; if we

---

1. *Father Melancholy's Daughter* (New York: William Morrow and Company, 1991), p. 52.

call it a memory at all, it must be one that belongs to a different order or kind of memory.

Our immediate response as to how such a case should be analyzed and understood has something to do with our court system and the way eyewitness testimony functions within it. We summon people as witnesses to testify as to what they have experienced *directly* regarding the case at hand: we do not ask them to repeat what they have merely *heard* about the case. (The police, of course, do solicit such comments when they investigate a crime.) And if their testimony should drift over into the transmission of what *others* have been saying, there is usually an objection raised. *Eye*witnesses are not supposed to present *hearsay*.

But there is also an ordinary sense of the term "remember" that does draw on "hearsay" or sources other than direct or primary experience. A teacher might ask a child in school: "Do you remember when World War I broke out?" In this case no one is supposing that the child was around to experience the outbreak of this war first-hand. Rather, the teacher had spent some time on World War I about a month earlier and is checking to see how much of the material has been retained. Now, many questions addressed to us regarding what we remember are of this character. And so it seems perfectly appropriate to suggest that people *remember* things that they have read about or been told about. It would be foolish to answer such questions by saying, "I know the answer to your question, but I don't *remember* it."

The basis of this distinction between two grades or degrees or modes of memory is a distinction between two types of experience: the one is direct, unmediated and presumably pure, whereas the other is contaminated by hearsay, the input of other people (who are not all to be trusted), and perhaps by other factors as well. But will this distinction hold up under careful scrutiny? If it should collapse, the understanding of memory with which many people operate in common sense thinking would have to be modified.

Direct experience. Suppose one man asks another whether he remembers the inauguration of President Kennedy. It was an unusual inauguration because of the weather — blustery conditions, with snow on the ground, which is rather unusual for Washington. If the second man answers that he *does* remember the Kennedy inauguration, what may we assume regarding his experience of this event? (i) That he attended the inauguration and heard the oath of office and the inaugural address with his own ears? Or (ii) that he was in Washington that day (with thousands of others) and took some part in the festivities? Or could it be (iii) that he watched the event on television at home (with no cold in the air or snow on the ground outside his home in southern California)? Or (iv) did he watch a television news special after coming home from work that evening, thereby catching the flavor and spirit of that memorable inauguration day? We cannot be sure from the answer he has given us. Some people might claim that only possibility (i) would count as remembering, but we could also be more generous in using the term "remember" and include all four as possibilities.

When Redeemer College held its very first graduation, there was not enough room in the auditorium to accommodate all the people who wanted to attend, and so the nearby gymnasium was used as an overflow room, where the events could be watched on closed-circuit television. Did those who sat in the gymnasium experience the graduation directly, and can they claim to "remember" it? To complicate things in this example, it should be realized that the procession wound its way through the gymnasium before entering the auditorium, giving the people in the overflow area a direct look at the graduates and the faculty in their colorful garb. Such an example indicates how difficult it is to apply the notion of direct experience in actual cases. Much of our experience is a mixture of what we took in with our own eyes and ears, on the one hand, with what was transmitted to us through various technological media or channels, on the other. And so our memories are often derived from a mixture of sources.

Consider another example. Suppose someone is being asked about an occasion when he listened to a sermon by a famous evangelist who is well known for the quality and power of his voice. If the questioner is more interested in the voice than the message, the person supplying the answers may soon be at a loss as to what to say. He may declare that he was greatly taken with the evangelist and his powerful message but then add, after a bit of reflection, that there was a first-rate sound system in the stadium where the evangelist's rally took place. Thus what he heard from his seat in the center-field bleachers, where he could only see the evangelist from the rear, was an electronically amplified version of his voice. What was he hearing *directly,* and what was he receiving on a *mediated* basis? This distinction is not easy to make. When we turn to singers and their amplification devices, the question becomes still more interesting, especially when we consider how much the quality of recorded music can be enhanced technologically, to the point that some people would much rather have the music to which they listen get processed by sound engineers before they listen to it. The success of technology is leading to a decline in the appeal of direct, first-hand experience.

What happens in a massive football stadium (where the field is so large that it is literally impossible for a spectator who stays in his seat to be in good viewing distance of all the action) also gives us reason to raise questions about the category of direct experience. Presumably one goes to the stadium where the game is actually being played in a quest for direct experience. Yet, some well-equipped stadiums now have massive instant replay screens, which spectators are encouraged to watch. It often happens in football that a spectator misses key aspects of an interesting play (perhaps he was taken in by the quarterback's effective fake) and therefore needs a replay to get a sense of what happened. In any case, a spectator is now invited to watch the actual play on the field and then the replay. Spectators are also known to bring radios to the game so that a play-by-play commentator can tell them what it is they are watching. Some even bring portable television sets. What happens when the spectator is asked one day, "Do you recall that famous play when So-and-so scooped up a fumble on his own two-yard line . . . ?" Let's say that the person being asked was at the game but missed the play; however, he got a good look at it on the replay screen, and also saw some replays on the sports report on television later that day. Can he claim to remember it? Presumably he does, and says, proudly, "Yes, I was there."

In view of what technology can offer in terms of watching sports events, one might even wonder why people still take the trouble to go to the stadium. Surely part of the answer is the excitement of being part of the crowd. And here an important feature of experience and memory comes into view: there is a *collective dimension* to be considered. To remember a key play in the championship game is not just to recall the sight of what transpired on the field; one should also recall how the crowd rose to its feet and roared. And crowd excitement calls for personal presence. A similar analysis could be offered regarding a presidential inauguration.

> *The optical instrument he sees through is so constructed that it encompasses only pre-determined fields of vision. Looking through it, he beholds only what he expected to see.*
>
> –Czeslaw Milosz[2]

The technology of experience. Since technology affects so many facets of our lives, we should not be surprised that it also shapes our experience in everyday ways that we often do not notice. The most obvious example is eyeglasses. Those who claim to prefer immediate, raw experience should be asked (in case they are near-sighted or have some other vision problem) whether they really would rather do without their glasses or contact lenses. That eyeglasses change the way we see the world is so obvious that one hardly thinks of it. It is obvious — and seemingly of no philosophical interest — because it changes our vision *for the better.* No one who is seriously short-sighted would want to do without the correction that lenses can bring. Indeed, when I drive a car, I am *required* to change my vision technologically by wearing my glasses.

Optical devices containing lenses put a frame around our visual field. Looking through a camera's viewfinder is an interesting experiment insofar as it reveals a certain effect on our experience: a child's toy telescope can give one the same effect. Technology not only enhances our vision, bringing distant things closer and magnifying small things so that we can see them; it also frames them in a way that

---

2. *The Captive Mind,* tr. Jane Zielonko (New York: Vintage Books, 1981), p. 31.

sharpens our focus. Presumably this is all to the good, but it does get us away from any ideal of pure, unmediated experience.

We could also consider technological changes of a different sort in our experience, e.g. those brought about by mood-altering drugs — and even diet. I do not mean to argue that all such changes are for the good: some of them pose obvious risks to human health. And many of the drugs and substances people use are addictive; hence one is allowed to take certain of them only under a doctor's supervision. But once we get past such drawbacks, it does appear that people seek to filter some of their experience through artificial "flavorings," as we might call them, such as the pleasant glow alcohol induces for many people. Insofar as we regard such changes in mood as legitimate, we are again backing away from the ideal of raw, unmediated experience.

Of a still different order is the way in which the technology called writing mediates the experience people have of one another ("intercourse" in the older sense of the term). Lovers are normally present to each other through speech, touch and overall closeness. But some lovers live many hundreds of miles apart. They will presumably have a good deal of telephone contact, but they will probably also write to one another. Before the telephone became part of our lives, such lovers had to carry on their romance almost entirely by correspondence. They were still present to one another in a way we could call intimate, but the medium of their contact with one another was writing. Some languages (e.g. French) are said to be more suitable for romance than others. In such a claim we see what a contribution technology and communications media can make to the shaping of actual human experience.

> *Beware of first-hand ideas! . . . Let your ideas be second-hand, and if possible tenth-hand, for then they will be far removed from the disturbing element — direct observation.*
>
> –E. M. Forster[3]

---

3. "The Machine Stops," in *The Collected Stories of E. M. Forster* (New York: Alfred A. Knopf, 1979), p. 182.

Guided experience. Nietzsche surely counts among those who advocate direct and unmediated experience. He advised us to "rebel against secondhand thought, secondhand learning and imitation," and he scoffed at people who "run through art galleries."[4] It appears that the people he had in mind are unwilling to take the time to plumb the depths of human experience: a summary, a reduction, will suffice for them. They know in advance what they are looking for. Weighed down with "historical education," they look at a painting to check whether the features they were told to find are indeed there, and then they scurry on to the next one.

Even though teachers make much of "primary sources" (which are themselves mediated by editors) and join Nietzsche in warning against "running through art galleries," much of their work amounts to guiding us in our experience — telling us what to find. Even a walk through the woods by a grade three class requires preparation: the children are told in advance what to look for. And tourists are famous for carrying guidebooks that tell them what the items of interest are in the town they are visiting.

What is important for our discussion here is that this element of guidance comes not only *before* we experience things but *afterward* as well. In many situations we are not left to ourselves to form permanent impressions in the form of memories. Examples can be drawn not just from cultural excursions but also from science laboratories. When something unusual happens in a science lab, the students may be left wondering just what it was and what bearing it has on the theories being explored, but the teacher usually has an answer ready. And so the students may have to revise their memories of what they saw — or *thought* they saw.

Yet we should not immediately denigrate or condemn such retroactive interaction between memory, on the one hand, and description that draws on language, theory, and perhaps pictures, on the other hand. For the most part, a verbal articulation of what was experienced

4. *On the Advantage and Disadvantage of History for Life* (Chapter 1, Note 5), pp. 64, 41.

serves to reinforce the experience: that's why a young child bursts into the house and interrupts his mother in order to tell her what he has just seen. Suppose someone on an all-day nature hike sees something very exciting and unusual shortly after he sets out. Because he is alone and encounters no other human being on the hike, he does not have an opportunity to tell anyone about it until many hours have passed. In all likelihood, the experience will by that point seem less exciting and significant than it would have been if he had had a companion with whom he could have discussed it throughout the day. Students also know that verbalizing something that was observed reinforces it as far as memory is concerned: that's one reason why teachers often give assignments in which students have to *describe* something they have watched, especially during excursions outside the classroom. But in the process of description, the memory that has been formed may well be altered, as the experience is fitted into the categories provided by the language being used.

Perhaps the most dramatic example of the fading away of experiences that are not discussed and regularized and semanticized is what happens to the mystic. A person undergoing unusual religious experiences (usually in isolation) often finds that he does not fit in well with his religious community when he returns to it. Part of the problem is that his experience sets him apart from the other members of the community, who did not share in it and probably have never felt anything like it. And if he goes out of his way to make his experience accessible to them, it tends to lose its unique character and to get remembered as something a bit different from what it was originally.[5] Hence it is often said that mystical experiences, as well as other very intense religious experiences, are "ineffable": nothing can be said about them.[6]

---

5. The extent to which mystical experience as actually remembered is dependent on what people (not just the participants, but others as well) later say about it and how they interpret it is explained by Agehananda Bharati in *The Light at the Center: Context and Pretext of Modern Mysticism* (Santa Barbara: Ross-Erikson, 1976), see especially pp. 67ff, 98, 189, 219.

6. This inability to express and communicate what was experienced or "seen" may well be a factor in the famous "silence of the Buddha." Buddhism, more

The extent to which our memories are affected by what people say to us afterwards has been studied especially in connection with eyewitness testimony, which plays a very important role in courtroom proceedings. Elizabeth Loftus observes: "During the time between an event and a witness's recollection of that event . . . the bits and pieces of information that were acquired through perception do not passively reside in memory waiting to be pulled out like fish from water. Rather, they are subject to numerous influences. External information . . . can intrude into the witness's memory, as can his own thoughts, and both can cause dramatic changes in his recollection." Eyewitnesses seem to crave harmony between their own recollections and the physical evidence relevant to the case at hand. Loftus writes: "What happens when a witness sees some event and later learns a new piece of information which conflicts with some aspect of what was previously seen? It appears that when possible many witnesses will compromise between what they have seen and what they have been told later on." Even the words that are used in questioning a witness to an automobile accident, for example, can influence his recollection markedly: substituting the word "smashed" for the word "hit" when questioning witnesses can have an effect on what is reported.[7]

When we consider non-verbal input, we again see the extent to which we allow our experiences and memories to be shaped by what others place before us. One reason why caricature, as used in cartoons, is so effective is that it shapes our actual perception and our retained sense of what people look like. Many a public figure winces at the drawings made of him by cartoonists, and must then suffer further as people tend to become more aware of the unusual features of his physical appearance and focus on them. The cartoonist (or comedian doing a voice impression) is a very perceptive individual who draws attention to aspects of a person's appearance and conduct which might not have been widely noticed before. There is a history to the way we see and remember people, for we are wide open to

---

than any other religious tradition, seems to regard the most significant of all teaching as beyond human utterance.

7. *Eyewitness Testimony* (Chapter 2, Note 34), pp. 86–7, 56, xii, 77–8, 96, 164.

influence. And so we see again that our experience of the world is by no means raw and unmediated.

> *Memories of the past are not memories of facts but memories of your imaginings of the facts.*
>
> –Philip Roth[8]

> *. . . I think up something when I'm alone, and afterward I can't remember whether I just thought it up or whether it really happened to me.*
>
> –Hans Erich Nossack[9]

A mixture of sources. Where does the material we find in memory come from? People like to think that it bears the impress of events we have experienced, and presumably it does. But memory is much more than just a deposit of what we have been through. Even if we view it largely as experiential residue, we must recognize, in view of the analyses presented earlier, that there is outside mediation and channeling in our experience. And so, if our experience is mediated, there must be factors other than direct, unmediated experience that help give rise to what we call our memories.

What we may find a bit difficult to accept is that there is *retroactive* interference with the contents of memory. (The normal human impulse is to trust the deliverances of one's own memory over what other people say; in old age many people surrender this reliance and become willing to be corrected.) I speak of this interference as retroactive and retrospective because it does not take place in random fashion: the way we remember earlier things has much to do with the way later things came out, which is why it is often said that history is a story told by the victors.

Perhaps an analogy can help here. A tape recording should be a

---

8. *The Facts: A Novelist's Autobiography* (New York: Farrar, Straus & Giroux, 1988), p. 8.

9. *Wait for November,* tr. Ruth Hein (New York: Fromm International Publishing Corporation, 1982), p. 78.

deposit or reflection of sounds in the air that were available to be registered at the time the tape was made. But we sometimes get strange results when we play a tape. We then conclude that there was interference with the tape after it was made: the presence of a magnet might account for the distortion we find. Now, a magnet normally spoils a tape in an indiscriminate way, but the retrospective interference we find in our memories is usually intelligent, even if, in some cases, it is to be regretted because it amounts to falsification. In dealing with this phenomenon I propose to use a somewhat neutral term and speak of the *editing* of our memories. My thesis, then, is that our memories are not inert but undergo a process of editing, whereby they are regularized, rendered more retainable, and reshaped with an eye to *subsequent* circumstances and events.

At this point we should consider the role of imagination once more. Memory and imagination do not stand for separate, watertight "compartments" or "faculties" of the brain having nothing to do with each other, nor should we think of them as two "departments" of the mind that are independent of one another in their operations. What we find in memory (or that which seems to present itself as memory) may well have roots in our capacity to make-believe. And sometime our pretending is so intense that we ourselves become confused. In short, there is no sure-fire way to determine whether something that seems to be a memory really is a memory, as opposed to being a product of imagination. One of the things that separates people who are judged to be sane from those who are thought to be insane is the ability to discriminate fairly effectively between memory and imagination: a sane person is able to persuade his fellow human beings that his memories really are memories.

Now, the mention of insanity may make the danger of which I speak seem more remote than it really is. Something with which I have difficulty in everyday life is confusion stemming from recollection of acts of imagination in which I see myself carrying out intentions. Let's suppose that I hear a piece of interesting and exciting news and form the intention to pass it on to my wife when I get home. I imagine myself telling her. Somehow the memory of that act of imagination later gets taken to be a memory of the actual deed. I wind up not telling her (since I have a sense of having already done so), but later on I claim that I do remember telling her. What I am calling memory in this example is really a recollection of something that took place in imagination alone. The mistake is easy to make.

> *Memories, she was well aware, depended on selectivity and structure as much as any drawing did.*
> –Mary Elsie Robertson[10]
>
> *If I have occasionally made use of some immaterial embellishments, this has only been in order to fill a gap caused by lack of memory.*
> –Rousseau[11]

Constructed memory. The recognition that memory draws on a number of sources, both within and beyond ourselves, opens room for the further realization that it is the product of a process of construction. I. M. L. Hunter writes: ". . . recall is often less a matter of literal reproduction than of the imaginative construction of fragmentary recall into a coherent whole. Normally the individual is unaware of this constructive characteristic of his remembering."[12]

The individual's lack of awareness of the process by which he edits and shapes his memories is an obstacle to the acceptance of any such understanding of memory. People do have some sense of what it feels like to engage in strenuous mental activity: for some, it results in headaches. The process of editing which memory undergoes, however, seems to be largely unconscious. Just as we sometimes discover that we have changed our mind about a certain matter, without remembering a time of reconsideration or being aware of just when the change came about, so we quietly reshape the way we remember things and thus alter the stories we may tell one day on the basis of the memories we are editing.

---

10. *Family Life* (New York: Atheneum, 1987), p. 57.

11. Opening page of the *Confessions* (London: J. M. Dent & Sons, 1931, translation credited to John Grant).

12. *Memory* (See Chapter 2, Note 9), p. 183.

The mental abilities that underlie our capacity to engage in historical thought, then, do not stem entirely from our talents as *storytellers;* the shaping of history is grounded even more deeply within us. Memory, in other words, is not an inert archive but seems to be a sort of continuing research project. According to Suzanne K. Langer, memory, understood in active terms, ". . . is the real maker of history — not recorded history, but the *sense of history* itself, the recognition of *the past* as a completely established (though not completely known) fabric of events, continuous in space and time, and causally connected throughout."[13] If this is indeed the case, a discussion of truth in history must begin not with narrative (important as it is in itself) but with memory as one of the major sources on which narrative must draw.

> *There you go again, Thea, creating memories out of whole cloth because you know you're safe. You know my memory's awful.*
>
> –Gail Godwin[14]

> *Recollection, I have found, is usually about half invention.*
>
> –Wallace Stegner[15]

Amnesia. As we probe the phenomenon of amnesia to see what it can teach us regarding memory, we should first note that it is not the total loss of memory that many uninformed people take it to be. People in a state of amnesia are still able to perform many actions that presuppose learning and intelligence: they have by no means forgotten or lost everything they once possessed in the way of knowledge.[16]

---

13. *Feeling and Form* (Chapter 2, Note 22), p. 263.

14. *A Mother and Two Daughters* (New York: Viking Press, 1982), p. 187.

15. *Crossing to Safety* (New York: Random House, 1987), p. 53.

16. In one of the classic cases of amnesia discussed by the psychologist William McDougall, a person suffering from amnesia proved capable of a feat requiring considerable presence of mind, but he could not recall just how he had done it: see Hunter, *Memory,* pp. 237–8.

Amnesia can better be understood as an inability to remember,[17] and it is usually remedied after a while — at least in part. Amnesia in this simpler episodic sense is a condition to which a great many people are subject on one occasion or another, especially when they grow older. A blow to the head may be sufficient to induce a small case of it. It can also result from the electric shock therapy sometimes used by psychiatrists. (The memories lost through such treatment do not all return.) People who suffer from amnesia of some such sort may envy so-called normal people, without realizing that memory for "normal" people is also episodic and discontinuous in many ways.

Amnesia is not properly understood unless we realize how much reconstructing and stitching together of webs of meaning goes on as we struggle to remember. Much of what we claim to remember is postulated by us as *likely to have happened* because it fits in well with what we do seem to remember fairly vividly. And we cannot simply inspect the contents of memory to see what is genuine recollection and what is mere postulation.

The amnesiac is often presented as someone who does not know who he is: in fiction we often encounter him without a wallet or personal identification.[18] But an amnesiac is not out of the woods, so to speak, if someone taps him on the shoulder and tells him his name. Even a more complete recital of personal data does not restore him to his sense of himself. His problem is that there is a strong synthetic element in memory which he seems to have lost. He lacks the capacity to bring together elements of information about himself (whether remembered or supplied by someone who knows him) into a coherent personal history which he can then remember as his own story and on which he can base an identity, a sense of who he is.

An identity or sense of self is never to be taken for granted. Our identity is tied up with our name and the degree to which we are recognized as So-and-so in the circles in which we move. We like to

---

17. When we try to remember something and fail, we do not assume that the fact or detail that escapes us is forever lost; the inability need not be permanent.

18. Evan Hunter's novel *Buddwing* (New York: Simon and Schuster, 1964) deals with such a case.

be called by name and to be recognized by people we meet; without such recognition and tokens of familiarity, we are weakened in terms of our sense of who we are, as many a person has discovered when he was far away from home in a place where no one knew him.

To remember who you are (in the fullest sense, which includes a moral aspect), you must be able to tie the elements you remember as making up your life into a coherent package. The continuity and coherence of that package is not a simple deliverance of memory but is always somewhat dependent on the constructive activity of the mind: selfhood is a moral-practical enterprise, a continuing task.

Past-life recall. The constructive activity of the mind in memory is crucial to assessing reports of past-life recall on the part of those who believe in reincarnation.[19] Although the Buddha was supposed to have been able to recall thousands of past lives, most human beings have difficulty remembering much of the earlier part of the life they are now living. Thus, whatever else past-life recall might be about, it does represent a claim to an astonishing feat of memory. And what is astonishing is not just the claim to be able to retain many details and episodes from past lives; it is also the ability to weave the threads of previous lives together into one story, which must in some sense be the story of a single self.[20]

Actual reincarnation reports generally do not make such a claim: what we usually encounter instead is a story wholly contained within the framework of a previous life, coupled with the claim "I used to be that person" (whatever that might mean). When a novelist turns to the theme of reincarnation, however, and tries to make a compelling story of it, a previous life and the current one get linked: typically, two or more people who had relationships in a previous life have relationships again in this life and gradually become aware that their former

19. These two seem to go together: if one does not believe in reincarnation, any memory phenomena that might appear to be instances of past-life recall will have to be explained and understood in some other way.

20. If no claim regarding a continuing selfhood is made, the doctrine of reincarnation becomes so weak that virtually anyone can affirm it. It is then the claim that when people die, new people are born — the stream of life flows on.

relationship is complicating their present one.[21] There are even psychotherapists who try to get people to regress to past lives in the conviction that relationships in the present are out of kilter because of something that happened in a past life. In Edgar Cayce's "readings" of past lives, we also come upon the theme of people involved in complications with people with whom they had relationships in one or more past lifetimes.

While there are some people who come to believe in reincarnation on the basis of past-life recall stories, many in the Western world look upon them with skepticism. The difficulty is that they are so hard to confirm. If we use vividness as a (quasi-Cartesian) criterion that guarantees authenticity in memories, they should be taken fairly seriously. But if vividness is not conclusive by itself, and if it is admitted that there is no foolproof way to determine whether material in consciousness stems from imagination (as opposed to memory), we are stuck with having to seek independent confirmation of past-life stories. And that confirmation is usually indecisive: the stories (many of which seem to be set in ancient Egypt) are not easily confirmed or disconfirmed, for there are various routes by which the material they contain might have entered a person's mind.

In good measure the reincarnation challenge is an ethical and spiritual one. The person who seems to recall previous lives must ask himself whether he wishes to seek a larger identity and selfhood that embraces previous lifetimes and serves to bring together all of these recollections, some of which are rather disjointed, into a larger whole. Christian teaching has traditionally suggested that we should *not* think in terms of previous lives, but if we consider the matter in purely rational terms there is in principle nothing to hinder me from imagining and believing that I am a reincarnation of Napoleon, for example. The question is: what are the implications for daily life? Does my being a reincarnation of Napoleon entitle me to special privileges? Will historians interview me?

A lesson from life under twentieth-century totalitarianism can help

---

21. For an example of such a story, see Barbara Erskine's novel *Lady of Hay* (New York: Delacorte Press, 1986).

us here. The history of brainwashing and of Stalinist oppression in our century has shown us that memories can be planted in people; that is to say, if people are placed under sufficient duress, they can be made to "remember" committing crimes they never actually committed. Perhaps we should simply conclude that someone with past-life recall experiences can be talked with and *either* challenged to integrate those recollections into his current existence (the reincarnation answer) *or* advised to set them aside as bizarre mental episodes (perhaps parallel to unusually vivid nightmares) that have no value for life as it is lived today. And if the latter route is chosen, the recall experiences may soon end, as they often do for people who have them in childhood but not in adulthood.

When we deal with past-life recall reports in the fashion I am suggesting, we are in effect treating the word "remember" as an honorific. Sometimes a person claims to *know* something, only to find that people challenge him and declare that his piece of alleged knowledge is only an opinion. For example, in the middle of the baseball season he might say that he *knows* that the New York Yankees will win the World Series this year — and then discover that people refuse to regard his certainty about this matter as a piece of knowledge. In similar fashion, someone who says he "remembers" having lived before and doing thus and such may well be contradicted and told that what he is reporting simply does not count as a memory. Just as philosophers often argue that the term "know" should be reserved for very strong instances of cognition, so the term "remember" might conceivably be reserved for plausible recollections that comport well with traces of the past, with accepted history, and with the recollections of other people.

Chapter 4

Sequence and Anachronism

One thing after another? We apply the term "story" to a faithful record of events as they actually happened, but we also use this word in application to fiction. The understanding that a story can be a piece of fiction, that is, something made up, is a potential problem for people committed to truth-telling as they talk about their own lives. How can we achieve a fundamental truthfulness in our narratives about ourselves?

Common sense is quick to respond with the suggestion that we ought to narrate events in the order in which they happened. To tell about things accurately, to explain to people "wie es eigentlich gewesen,"[1] would presumably mean sticking to the *chronological order.* And so we must consider the suggestion that sticking to temporal sequence is the main guarantee of truthfulness in history.

Now, there are various reasons why sequence cannot serve as the key to truthfulness in history. It turns out that in many cases it is impossible to be absolutely sequential, and in quite a few others it would be exceedingly cumbersome to adhere strictly to sequence. One of the main reasons why sequentiality cannot be strictly carried through when we tell about the past is that historical narratives and even shorter statements about the past are shot through with anachronism.[2] But even apart from the problem of anachronism (which will be discussed later in this chapter), there are various reasons why sequence breaks

---

1. How it really was. This famous phrase stems from the German historian Leopold von Ranke (1795–1886). His actual view of the writing of history was somewhat more sophisticated than one might think from considering this phrase.

2. See the Glossary.

down in our narratives, even if we tend not to notice many of the departures from strict sequentiality.

This is not to say that sequence has no place at all — far from it. Within a portion of a narrative, events will generally unfold in chronological order. Yet even in small narratives segments, there are often gaps to be noted. Observing sequence in the strict sense would mean filling those gaps, but a narrator telling his own story in a fashion accepted as ordinary by common-sense consciousness thinks nothing of leaving them unfilled. We do not feel constrained, then, to tell things one after the other; apparently it suffices to tell only *some* of them — and then to arrange those items in chronological order unless there is some good reason *not* to do so. Indeed, it would not be possible to tell absolutely *everything.*

> *Past and future are, as it were, gathered up in the present rather than distinguished from the present in the immediate experience of living.*
>
> –John Macmurray[3]

Flashbacks. One standard violation of sequentiality is a phenomenon we often encounter in movies and novels: the pattern of moving back and forth in time. Sometimes a novel or movie will begin with a striking scene or encounter, and then move to the protagonist's past, only to work forward toward the opening scene again. In other cases flashbacks to earlier times are sprinkled throughout the narrative. They can be used skilfully to great artistic advantage; yet sometimes they are overused, with the result that readers and viewers begin to find them irritating or tiresome. There is still something to be said for a story in which time moves in one direction only. Some of the best stories, plays and movies have all their action unfold in a single day.[4]

There are also flashbacks of a less obvious sort sprinkled throughout many stories. Those flashbacks are bits of material from an earlier time entering into the constitution of the situation and action as it develops; in other words, we must consider them part of the con-

---

3. *Interpreting the Universe*, second edition (London: Faber & Faber, 1936), p. 29.

4. Aristotle laid down the requirement that the action in a drama should cover no more than a single day: hence we speak of "unity of time."

sciousness of the person(s) acting in the narrative, for without them the unfolding story would not make sense. Consider the following narrative, in which Harry, a young businessman, is giving his best friend a narrative explanation of why he did not ask Mr. Higgins for a business loan, as he had planned to do.

> ◘ I asked for an appointment, but I didn't tell him what I wanted to discuss. He said to be there first thing in the morning. So at eight o'clock I was driving down Main Street — nervous, as you might expect. It's been twelve years since his daughter Sally decided she didn't want to marry me after all, and I still think her family was largely behind our break-up. I could always sense what they really thought: I wasn't good enough for her.
>
> Anyway, here I was cruising down Main Street when I caught a glimpse of my old friend Luke Walker strolling down the sidewalk. Now, I hadn't seen Luke in at least fourteen years, and I hadn't heard anything about him visiting town again. I was so surprised to see him that I immediately slowed down and turned my head to the side for a better look. I wondered whether I should park the car quickly and run after him, but I had this appointment to keep. Anyway, while I was trying to get a better look at Luke — if it really was Luke — the driver in front of me braked suddenly. I turned my attention back to the road, but by the time I put on my brakes, it was too late: I hit him! I had almost managed to stop, but I hit him and caused some damage to his car.
>
> Well, as you can imagine, the driver jumped out and waved his arms and shouted. Who do you suppose it was? Mr. Higgins, on his way to his office to meet me! After that I didn't have the nerve the bring up the loan. He finally calmed down and asked me what we were supposed to meet about, and I blurted out that I wanted his advice about a tax matter. And he said that because of the accident he wouldn't be able to make time for me today — he'd have to see about getting his car repaired.

The action in this little narrative is fairly unified and sequential: Harry makes an appointment with Mr. Higgins and sets out for his office the next day but "runs into" Mr. Higgins on the street, with the result that his original intention comes to nought. It all happens within less than a day. Yet this story contains two significant flashbacks in the form of bits of information about past relationships and states of affairs, both of which are relevant for understanding how the action in the story's narrative present unfolds. (i) Harry's nervousness is rooted in the break-up of his engagement to Mr. Higgins's daughter Sally twelve years earlier: the Higgins family apparently put on pressure that led to the termination of the engagement. (ii) Harry's relationship with Luke Walker fourteen years before was sufficiently exciting and important to him that a glimpse of a man who appeared to be Luke caused him to divert his attention from the traffic on the road, and this was the immediate cause of the accident.

Now, we surely would not require that this story, if it is to be told in a truthful way, should first deal with the relationship between Harry and Luke Walker (fourteen years ago) and then go on to the break-up of the relationship between Harry and Sally (twelve years ago), before proceeding to Harry and Mr. Higgins in the present. If such a requirement were laid down for larger narratives, they would become difficult to follow. The placement of the flashback in the narrative we are considering is determined by narrative relevance: to understand something in the present, we need to be told (or perhaps reminded) about something in the past.

This is not to say that all flashbacks are narratively necessary. But they do add to comprehension in most cases, even if some of them appear designed to add complexity and keep the reader guessing. Perhaps there is a tension between the extent to which a story is comprehensible and the extent to which it is a faithful account of what happened. In some cases the truth is just too complex to be easily understood; it may have to be simplified. ("What it basically boils down to is . . . .") This drive toward the comprehensible is also a factor in memory, as we saw: we tend to compress, reduce, simplify.

Intersections. Another reason why strict sequentiality cannot be observed in narration is that in many cases parallel sets of events unfolding at the same time intersect and clash. If the events that make up a narrative were always as neatly related to one another in time as dominoes that fall over one by one, each knocking down the domino ahead of it, we might be able to remain strictly sequential. But the stories we tell about real life are often messier and harder to observe than a row of falling dominoes. Consider the following third-person narrative, in which Jeffrey is seeking just the right circumstances to ask his girlfriend Linda to become his wife. He senses that their relationship is cooling off and fears that she might say no.

> ◘ Jeffrey ushered Linda to a reserved table in their favorite restaurant, where they had eaten on a few other special occasions. Over dessert he finally began to inch toward the big question. He would have made his move a bit earlier, but there was a rowdy group at the next table. He was uneasy about the rowdies, and so he waited, hoping they would leave, but they seemed in no hurry. Finally he felt he had to steam ahead or lose the occasion altogether.
>
> Just when Linda began to sense where Jeffrey was headed, the quarreling at the next table escalated into shoving, and then a punch was thrown. The man who took the punch lurched backward in Jeffrey's direction. Jeffrey managed to rise quickly and scramble out of the way, but in the process he tipped the table and sent Linda's dessert to the floor with a clatter that drew the attention of the entire restaurant. Red-faced, he said to Linda, "Let's get out of here."
>
> Once he was out on the street, he began apologizing to Linda for not saying something to the rowdy men who broke up their intimate little dinner. It bothered him that Linda had seen him in a bad light, and he decided he would bring up the subject of marriage some other time.

In this case, action taking place within one group of people spills over into the action in another group that just happens to be nearby, interrupting what was underway in the latter group and destroying what was supposed to be a special occasion. Jeffrey and Linda are free to go to another restaurant, order some more dessert, and pick up where they left off, but that's not generally what people do in such circumstances. The mood has been shattered by an intrusion from without.

A sequential narrator could not very well be expected to tell about the quarrel at the rowdy table before embarking on the story of Jeffrey and Linda. If chronological order is thought to be important here, the best thing is to shift back and forth between the two tables, which is what many fiction writers would do in such a case.

The temporal coincidence of separate streams of action, whereby one stream plays a causal role in the other, is not a flaw in a narrative but is rather a feature which storytellers draw on heavily — perhaps too heavily in certain stories. While some narrators and novelists move the story ahead mainly by what they find in the character of the people they are telling about, others love to bring in coincidences — other actions that happen unexpectedly[5] and have the effect of influencing or advancing or perhaps blocking the main action in the story. Such a narrative technique is popular, for people love a surprise. But to use it properly requires that sequentiality be violated.

Anticipations. The last reason why strict sequence cannot always be maintained in a narrative is that human actions themselves transcend time in the sense that they anticipate reactions and are often planned with an eye to the anticipated (future) reaction. The reaction comes later in time than the action, but it is present before the action in the sense that it has a hand in shaping the action. This can be demonstrated easily in a back-and-forth game like tennis: the part of the court to which one hits the ball depends in part on what one expects one's opponent to (be able to) do in terms of returning the ball: a tennis player may exploit his opponent's weak backhand, for example. In football we see a similar pattern, this time rendered more complex because it involves quite a number of players on each side.

If we use football as our example, it is easy to see that there are not only anticipations at work in human action but also anticipations of anticipations. For example, a certain formation which the offensive team uses when it is about to put the ball into play will usually bring about a certain reaction in terms of a defensive formation. But the defensive team will try to anticipate the play (and the formation from which it is run) well before the snap of the ball. Yet the offensive

5. The action in such a case is unexpected in relation to the primary events and characters in the narrative. This is not to say that it could not have been foreseen and expected within the context of the series of events in which the secondary agents are embedded. The unexpectedness is mainly a matter of temporal and spatial contiguity.

team in turn anticipates the defensive reaction to the overall situation in the game and to its offensive formation and therefore may come with a different formation for running one of its standard plays in an effort to keep the defense confused and off balance. In principle there is no limit to the number of levels of anticipation and response. The same point can be made by reference to strategies in chess. There are also such dynamics at work in tennis: if your opponent expects you to play to his weak backhand, you might surprise him at a key juncture in a game by challenging his stronger forehand.

In the world of human interaction, then, we must not think in terms of actions which neatly follow upon one another in time, nor can we say that one (completed) action invites and makes possible an action in response; rather, the future (expected) action enters into the nature and constitution of the action being undertaken in the present. In simple terms, we could say that actions and reactions interpenetrate each other.

Marital disputes also provides illustrations of these patterns. Consider the following story, which is told by an old-fashioned omniscient narrator.

> ◘ Gordon came home half drunk, as he had done so often. His wife Gloria, suspecting that he was seeing another woman, screamed and carried on. Gordon decided to call a halt to her antics, and so he declared he was moving out that very night and would seek a divorce. He assumed that Gloria did not have the courage to face life on her own after eighteen years of marriage; she would be frightened into submission.
>
> He was in for a surprise: she shoved some of his clothes into a suitcase and pushed him out the front door, locking it behind him. What Gordon did not know was that she herself wanted out of the marriage by this point and was looking for an opportunity to place herself clearly in the position of the aggrieved party so that she could win a generous settlement in the divorce court.
>
> Ten minutes later, Gloria heard some noise downstairs: it was Gordon, who had let himself back in. By now he sensed what she might have in mind and realized that it would be a tactical error to leave the house under such circumstances. He now began to play the role of the repentant husband.

In Gloria's conduct we have a case of anticipation of anticipation. Gordon's demeanor on coming home was clearly expected on her part. She anticipated that he thought she would put up with it, and so she threw him off with a furious reaction. This led to an ill-conceived strategy on his part, which she had also anticipated — that he would threaten her with divorce. Thus her initial response to the state in which he arrived home anticipated not just his reaction to her but her further reaction to his reaction, which would put her in the driver's seat for the divorce negotiations for which she was secretly preparing. In the light of these various levels of anticipation that so often are operative in human life, we simply cannot maintain that human actions follow upon one another in a strict and discrete temporal sequence.

> *He had played the fool with one of the most sacred ideas still left to a world where the notion of sacrosanctity had become abhorrent — the idea of Time. He had dared to be of a time not his own.*
>
> –Robertson Davies[6]

Can anachronism be avoided? The term has something of a negative ring to it, and so people generally assume that anachronism *should* be avoided wherever possible. What comes to mind first of all is anachronism in connection with conduct and action: a woman might act and dress like a 1920s flapper in the 1990s and be chided by her friends. To avoid anachronism is to move with the times when it comes to culture, style, technology. Some used bookstores rely on quaint, old-fashioned cash registers to ring up sales and make no use of calculators and computers as they conduct their business. They are free to do so, but many people would regard such deliberate anachronism as wrong-headed and probably costly in business terms.

The violation of time can also take place in reverse order. If a writer had someone using a telephone in a story which was set in the American Civil War (1861–65), he would be guilty of an anachronistic

6. *The Lyre of Orpheus* (Toronto: Macmillan of Canada, 1988), pp. 466–7.

error, for the telephone was not invented until the 1870s. A classic and most interesting example of an anachronism occurs in *What's Bred in the Bone,* a novel by Robertson Davies, in which a budding young art expert exposes a painting as a forgery on the basis of a detail he spots in it. The painting, which is supposed to date from the early fifteenth century, includes a monkey hanging by its tail. The monkey in question is a native of the New World and therefore could not have been known to the European painter who was alleged to have produced the painting. Hence the painting was bogus.[7] Documents that are alleged to be old are tested for authenticity by seeking chemical anachronisms: if the ink or the paper contain chemical ingredients developed after the date from which the document supposedly stems, it must be a forgery.

But the more important meaning of the term "anachronism" has to do with narratives. Many of our narratives — and the individual sentences of which they are made up up — contain anachronisms. Indeed, they are so deeply embedded in narrative discourse that we are generally not aware of them as we use them, although we do recognize them when they are pointed out to us. Consider the following short narrative about the Bible's King David, which is saturated with anachronism.

> ◘ Israel's king learned the art of kingship while tending his father's sheep near Bethlehem. His courage grew, and so did his patience. And when a sheep wandered away one day, he did not rest until he had found it and returned it to the fold.

To begin with, the boy tending sheep is identified as *Israel's king.* He was no king while he was out in the fields (although he later became one), nor did he then know he would one day become king.[8] In the second sentence, qualities he needed and possessed as king are spoken of: while caring for his father's sheep, David was in training to become king and learned the kingly virtues. Yet he had no way of knowing

7. See *What's Bred in the Bone* (Toronto: Macmillan of Canada, 1985), pp. 339–55.

8. I am placing this narrative before David's anointing by Samuel: see I Samuel 16:1–13.

that this was the larger (divine and providential) purpose of his experiences out in the field with the sheep. The final sentence is anachronistic in another sense: it clearly reminds us of what Jesus says about the good shepherd (see John 10:1–18), and so we project the dedication of that shepherd back onto David, who, presumably, was also concerned about each of his sheep. Indeed, David's dedication as a shepherd is intended to teach Bible readers something about certain facets of Christ's relationship to his "flock."[9]

This little narrative is instructive not just because it exemplifies anachronism in general, since it reads future, completed states of affairs back into earlier states of affairs (e.g. by calling the boy in the field "Israel's king"), but especially because it brings in a key dimension for understanding anachronism in a Christian context, namely, *God's* involvement in human actions. David may not have known that his experiences out in the field would stand him in good stead one day when he became Israel's king, but God did: what happened to David during his shepherd days was part of God's program of education for him.

The Bible is a thoroughly anachronistic book in that it regularly invites us to read earlier passages and themes in the light of later ones. David as shepherd in the Old Testament sheds light on the great and good shepherd who stands revealed in the New Testament, but the New Testament completion of the shepherd theme is also de-

9. Joseph Heller makes fun of anachronism as a concern in literature in his novel *God Knows* (New York: Alfred A. Knopf, 1987), which is made up of musings by King David just before his death. David refers to future technological and cultural artifacts and once in a while reminds himself that they haven't been invented or developed yet. The other characters in the book also get in on the act: Heller even has Bathsheba asking David to get her a word processor! Moreover, David comments on the way people in the future dealt with him, e.g. Michelangelo's statue of David. On the other hand, there are limits to David's knowledge of the future. All throughout the book he faces the issue who should succeed him as king — Adonijah or Solomon. Bathsheba keeps after him to declare her son Solomon his successor. David doesn't really want Solomon to take his place and doesn't know how things will come out. In the end, Adonijah overreaches himself, and the crown goes to Solomon. The story would not work if David knew in advance how this struggle was to be resolved.

signed to color and heighten our appreciation of Old Testament references to shepherds.

I could also point to Moses as an example. In Hebrews 11 we read: "By faith Moses, when he was grown up, refused to be called the son of Pharaoh's daughter, choosing rather to share ill-treatment with the people of God than to enjoy the fleeting pleasures of sin. He considered abuse suffered for the Christ greater wealth than the treasures of Egypt, for he looked to the reward" (vs. 24–6). The primary account of the events spoken of here is to be found in Exodus 2, where we read nothing about Christ or about lofty motives on the part of Moses.[10] Yet we do know from the Old Testament record that his flight from Egypt was an important episode in the history of redemption. The passage in Hebrews authorizes us to speak of *Christ* in connection with these events; perhaps we could say that it elevates and dignifies the story.[11]

The Bible, then, invites us to read Old Testament beginnings in the light of New Testament fulfillments. This pattern has taught Christians how to interpret individual lives — especially their own — as they talk about God's providence and "leading." There is usually more going on in a person's life than he realizes from day to day. Narrative anachronism is bound up with Christian consciousness at a very deep level.

It is perfectly true, as philosophers say,
that life must be understood backwards.

---

10. There is also an account of this episode in Acts 7:23–9, where we find no mention of Moses consciously suffering for Christ either. Yet this passage does include the notion of a transcendent conception of the situation on the part of Moses: "He supposed that his brethren understood that God was giving them deliverance by his hand . . ." (vs. 25).

11. This way of dealing with the Moses episode is typical of a certain strand of Reformed preaching and Bible reading, in which the unity of the Bible is stressed: see my book *Reading the Bible as History* (Burlington, Ont.: G. R. Welch, 1980). Such an approach to the stories in the Bible is carried through consistently in S. G. De Graaf's *Promise and Deliverance* (four volumes), tr. H. Evan Runner and Elisabeth Wichers Runner (St. Catharines: Paideia Press, 1977–81). The fulfillment of salvation history in Christ is used to (re)shape the telling of the Old Testament stories.

But they forget the other proposition, that it must be lived forwards.
–Kierkegaard[12]

Prospective and retrospective discourse. Anachronism is just as persistent in statements that have reference to the future as in statements about the past.[13] Yet we seem more tolerant of future-directed anachronism; indeed, it would be hard to imagine how we could get along without it. Consider an example: suppose I am driving to work and someone with me in the car asks me what I am doing. "Driving to work," I respond, but he says, "No, you are driving down Main Street." If I reply that I need to drive down Main Street in order to get to work, am I guilty of anachronism? Or suppose I tell my argumentative friend one day that I am painting the living room and he insists that I am (merely) moving furniture around. If I maintain that I must move furniture in order to be able to paint, am I guilty of anachronism, of projecting a completed action (the future) back into my description of (present) actions or subactions?

Similar arguments could be adduced against many statements made by politicians. A politician might claim to be improving the lot of teenagers when he is in the process of trying to raise the minimum wage. His action (if completed) *might* have the effect of improving the lot of teenagers, but then again it might not: informed opinion is divided on this score. Politicians often claim they are bettering the lot of some disadvantaged group whereas a stricter and politically neutral description of their action would confine itself to stating that they are changing the social, political and economic conditions.

The pattern we see here is that we tend to respond to "What are you doing?" questions with *achievement* verbs, that is, verbs stating that something worthwhile has been accomplished. For the most part, our use of such verbs in everyday language is necessary and justifiable. First of all, we need them as shorthand: we rarely have time to spell

12. Journals for 1843, in *The Journals of Kierkegaard, 1834–1854,* ed. and tr. Alexander Dru (London: Fontana Books, 1958), p. 89.

13. The term is generally used in reference to the past. I am using it in an expanded sense here.

things out in the kind of detail that would be required if we are to avoid anachronism in our future-oriented statements. In the painting example, the non-shorthand version would have to go something like this: "I am moving furniture around. Then I intend to get out my painting supplies and open the can of paint I bought last week. I will carefully apply paint to the walls, using brushes and rollers. When I'm all done, I'll ask my wife what she thinks of the work. Then I'll find out whether all this effort counts as 'painting the living room.' She might instead call it 'making a mess.'" (Note that even this version does not spell out all the subactions.) If we substitute a small child for the adult wielding the paintbrush, there is a high probability that the end result will be deemed a "mess," even if the child (perhaps inspired by his father's example) declares that he is "painting the living room."

The second reason why we need these achievement verbs is that if we were denied their use, we could not communicate effectively; that is to say, would not be able to say the things we really need to say to one another on occasion. Just imagine how difficult life would be if we were *never* allowed to declare our intentions and plans by means of achievement verbs. There are some people whose diffidence shows us what problems would result, and they are people with whom we tend to become exasperated. We see them busy with some project or other and ask them what they are doing. Not wishing to jump the gun, they evade our query by naming the subactions in which they are engaged. Suppose I walk by a house and see a great deal of digging going on, with earth heaped up on one side of the yard. I ask the workers what they are doing, and they tell me they are digging a hole alongside and under the house. (That much I could see for myself, of course.) In actual fact, the *plan* is to move the entire house over some twenty feet or so.[14] Because this is a rather unusual enterprise whose success cannot be guaranteed at the outset, the reluctance to use achievement verbs is understandable. Writers often manifest the same reluctance when people ask them whether they are "writing a book."

---

14. This example is a true story: such a project is underway in my neighborhood at the time of writing.

Whether the work underway will eventually result in a published book (as opposed to languishing unread in the writer's filing cabinet) will depend in part on other people — on what they think of the material, on their willingness to back it financially and promote it. Many a writer is reluctant to say (unless he already has a publishing contract in hand) that he is "writing a book."

The solution to these difficulties is not to *banish* achievement verbs as part of our way of describing the actions we are engaged in. Instead we should add the word "try" to some of our statements. "What are you doing?" someone asks. You might reply: "I am *trying* to paint the living room, *trying* to improve the lot of the poor, *trying* to move my house, *trying* to write a book."

The conclusion to be drawn is that there are two very important types of anachronistic discourse in wide use in human life. The one kind I call "prospective" because it defines and clarifies present action in the light of the anticipated future completion of that action. The other kind I call "retrospective" because its orientation is toward the past. The latter kind is the main object of interest in this essay dealing with our narratives about the past. But it helps to see that the former, i.e. prospective discourse, pervades much of what we say on a day-to-day basis.

Once we realize how much our day-to-day commentary on our own actions violates strict chronology, we will not be so quick to condemn anachronism in historical discourse, nor will we conclude that any statements about the past which could be judged anachronistic must be deemed misleading and must therefore be rejected as untrue, or as unworthy of our belief. Instead we will learn that we ought to be asking which features in discourse about the past we can appeal to and maximize in order to heighten the quality and truthfulness of our account of the past.

> *Most of life is a mess and a muddle; all chance and luck. Perhaps you can look back years later and say, well,* this *happened, or* that, *and then it will all fall into place, cause and*

effect, straightforward and logical. That is the business of fiction, to put things in order.
–Nina Bawden[15]

Completed experience. At this point the great significance of literature — especially the novel — for understanding memory and narrative in their everyday functions should begin to become apparent. I do not wish to offer an overly expansive generalization to the effect that all people in Western society today draw on literature in such-and-such a manner to structure their consciousness of their personal past. Not all of them read, or are even *able* to read — to say nothing of reading novels that present us with sophisticated narrative structures. What I shall observe will thus apply much more to some human beings than to others.

Various thinkers have supposed that "primitive" or early man was less introspective, talkative, and concerned with his own inwardness than man in the late twentieth century. I suspect they are right: we are somewhat less sober, objective and matter-of-fact about ourselves than most earlier human beings were. We rely more on a continuing narrative about ourselves (and perhaps less on socially assigned roles and places) to develop and maintain a sense of who we are. In other words, the truth about ourselves is not just "found" by us when we look back at our own experience; it is also "made," constructed, (morally) projected. But it is important to remember that this applies more to people today than to the so-called primitives.

Suzanne K. Langer helps us understand the appeal of literature when she explains that it presents events "in the mode of completed experience." This suggests that some experience is incomplete. She maintains: "Life is incomplete unless we give it form." How can we make life whole, complete? "The power to conceive of life as a single span enables one also to think of its conduct as a single undertaking, and of a person as a unified and developed being, a personality."[16] The unification of life in memory and narrative, the interlacing of the

15. *Walking Naked* (New York: St. Martin's Press, 1981), p. 15.
16. *Feeling and Form* (Chapter 2, Note 22), pp. 164, 400, 332.

threads, is to many of us such a fascinating process that we seek its fictionalized version in the novel, and then are affected by the novel in terms of how conceive of ourselves and nourish our consciousness of our personal past.

There is in people a hunger for interpretation, for the meaning of events. Completely random events can attract our attention, but they do not hold it for long. We are driven to assign meanings, and we are more apt to assign them prematurely, that is, to *over*interpret what goes on, than to *under*interpret. Indeed, what we call paranoia can be understood in part as the overinterpretation of social events. The person who enjoys sound mental health is one who has the ability to *resist* excessive interpretation. And so, if we err, it is usually in the direction of a premature completion of experience, of an imagined greater wholeness and fullness in our lives than close examination would reveal.[17] The day-to-day fragmentation and disorientation that really is present is hard to acknowledge.

Moral language. Common-sense thinking operates with a distinction between literary and moral language. It assumes that much literary language conveys to us things that are "made up"; therefore it cannot partake of truth, however amusing or pleasing people may find it. As for moral language, common sense recognizes that it cannot give us a statement of truth in the ordinary sense of presenting that which is the case. (Presumably historical and scientific discourse present real states of affairs.) Yet it is thought that moral language at least *aims at* the true, the right, the good. However, in the light of the consideration of anachronism, we see that the strict disjunction between these two which common sense wishes to make needs to be muted. And this muting of the difference is recognized by various writers, critics and thinkers who see a *moral* component in fiction and literature.

In considering these issues, it is helpful to note an older meaning

17. The classic idea of confession, i.e. God "looking over our shoulder" as we talk about ourselves, is a sobering prospect. "Thou desirest truth in the inward being," says the psalmist (51:6). He does not assure us that God usually *finds* truth when he looks within us.

of the term "moral," i.e. having to do with action, with the practice of human life. Bearing this meaning in mind, we might say that moral philosophy is not simply an investigation of rights and wrongs but also an analysis of how people act and can best realize their goals. As such, it intersects with psychology and various behavioral sciences.

With this broader meaning of the term "moral" behind us, we can now affirm that anachronism has a place in narrative and memory because of our *moral* needs. If anachronism in the form of prospective discourse were banished from our language and thought processes, we would become less efficient in the domain of action. Well then, if anachronism can be justified as a moral necessity, if we conclude that we cannot get along without *prospective* discourse, this justification should lend weight to the suggestion that anachronism need not be the enemy of truth when it comes to *retrospective* discourse either — however much we may fear the retrospective falsification of people's life stories.

The striking success of some totalitarians, e.g. the Stalinists, when it comes to falsifying the past, has led people to fear the notion of a continuing reinterpretation of the past on the basis of current successes and interests. Perhaps we should say not that history is written by the victors (which tends to leave room for Stalinist manipulation) but that it is written by *survivors.*[18] The definitive memoirs of World War II, which shaped our historical understanding of that conflict, were written not by President Roosevelt, who died before the war was over, but by Prime Minister Churchill, who lived on for many years after the war.

We also need anachronism in our moral language to make it possible for us to take responsibility for our past, for things we *have* done but should *not* have, and also for things that we should have done but failed to. Unless we allow ourselves to link up the present and the

---

18. This realization also helps us understand the great imperative that lived in the hearts of some of the people in Nazi death camps during World War II: they *had* to survive in order to *tell their story* to posterity. See, for example, Jean-François Steiner's account of what happened at Treblinka in his book *Treblinka,* tr. Helen Weaver (New York: Simon and Schuster, 1967).

past anachronistically, we will not be able to think clearly about our moral commitments or speak of them to others. Consider the following narrative in which omission looms large.

> ◘ Perhaps I was too immature to get married — I don't know. But I was away from home off and on for the better part of the first three years we were married, leaving Judy to cope with the twins. I sent money to her regularly, and most of the time she was just fine. But when Bobby suddenly got sick that night (I had left a day earlier for a construction job up north), Judy was without transportation, and sixty miles from the nearest town and doctor. Maybe she could have done more to rouse the neighbors to action — I don't know. But what I do know is that I had run out on her and on Bobby. I'm the reason he lost his hearing. If I had been where I was supposed to be, he'd be entirely normal today. But I abandoned him. I'll never be able to make it up to him, but at least he knows that I'll always be there for him.

In this narrative a man offers a retrospective interpretation of his own actions. He recognizes that he did not live up to his responsibilities. His absence from his family when his son suddenly became ill is interpreted by him as a case of abandonment, although at the time he went away he did not know that the child would become ill. Yet, looking back he realizes and admits that difficulties in his relationship with his wife were driving him to seek work far from home.

If we were to enforce strict rules against anachronism in narratives, we would have to dismiss this moral narrative as flawed. But instead we want to commend this guilty father for facing up to his responsibilities and committing himself to a new start with his family, even if we would probably also say that he has exaggerated his guilt in relation to his son's illness and subsequent loss of hearing, for he had no reason to suspect that the boy might suddenly become ill. This type of retrospective assumption of responsibility for what we did *not* do — but should have done — is surely part of what is involved in living one's life as a moral being.

The conclusion to be drawn, then, is that moral language does not allow us to live in a razor-thin present any more than literary language does. In other words, we do not live our lives or plan them or tell about them in terms of separate, discrete temporal units that neatly succeed one another. Moral language is largely future-directed; it allows the future to shape our thinking as it talks about the present, but it also includes elements of retrospective discourse insofar as it looks to the past for grounds for our commitments and for a larger framework within which to place them. A moral human being lives in the past and the future even while taking the present with the utmost moral seriousness.

Chapter 5

Beginnings and Endings

Whether I shall turn out to be the hero of my own life, or whether that station will be held by anybody else, these pages must show.
–Dickens[1]

Do we live stories? There are many people who suppose that a person's life is a series of stories. When he tells stories about his own life, he should do no more than report exactly what happened — there is no room for embellishment. Such people maintain that it is possible for us to be objective about the stories that make up our life just as we can objectively and honestly report what the digital meter at the service station indicates about the amount of gasoline we have pumped into our car. But there are others who point to an element of *discovery* when they tell stories about themselves. Some people enjoy telling stories about their own lives because they don't quite know how the stories will come out or end. And those same people are sometimes surprised at the opinions they express to friends.

Philosophers are divided on the question whether we actually *live* the stories we tell about ourselves. Some claim to find a narrative structure within experience itself. David Carr maintains that events are "narrative in character" and argues that narrative is "our primary way of organizing and giving coherence to our experience." The narrative structure of human experience and action does not just apply over longer stretches of time but can also be ascertained in "the most

1. The opening words of *The Personal History of David Copperfield.*

elementary experience or action."[2] Alasdair MacIntyre is another philosopher who believes we *live* the stories we tell before we tell them (at least, if we are truthful about our lives). Man, he assures us, is "essentially a story-telling animal," presumably because he is a story-*making* animal. Important for understanding MacIntyre's position is his emphasis on "intelligibility," which he identifies as "the conceptual connecting link between the notion of action and that of narrative." Thus he attributes "a certain teleological character" to our "lived narratives"; in other words, he finds considerable intentionality — or perhaps old-fashioned deliberateness — in human life.[3]

But there are also philosophers who maintain that stories do *not* arise directly and wholly out of experience; rather, they are made up afterward in an effort to make sense of experience and retain some of it for future reference and instruction. Among them is Louis O. Mink, who observes: "Stories are not lived but told. Life has no beginnings, middles, or ends; there are meetings, but the start of an affair belongs to the story we tell ourselves later, and there are partings, but final partings only in the story."[4] Hayden White is of the same persuasion: "No one and nothing *lives* a story."[5]

> *I didn't tell you because I knew you'd find it too interesting and then you'd force me to give you my story. . . . I wanted to mull it over myself.*
>
> –Gail Godwin[6]

---

2. See *Time, Narrative and History* (Bloomington and Indianapolis: Indiana University Press, 1986), pp. 4-5, 16, 46, 61ff, 65ff, 88.

3. See *After Virtue: A Study in Moral Theory,* second edition (Notre Dame: University of Notre Dame Press, 1984), pp. 211–16.

4. "History and Fiction as Modes of Comprehension," in *Historical Understanding,* a collection of Mink's writings edited by Brian Fay, Eugene O. Golob and Richard T. Vann (Ithaca and London: Cornell University Press, 1987), p. 60.

5. "Historicism, History, and the Figurative Imagination," in *History and Theory,* Vol. 14, No. 4 (1975), Beiheft 14, p. 59. On White's views see also *The Content of the Form: Narrative Discourse and Historical Representation* (Baltimore and London: Johns Hopkins University Press, 1987) and *Tropics of Discourse: Essays in Cultural Criticism* (Johns Hopkins, 1978) and *Metahistory: The Historical Imagination in Nineteenth-Century Europe* (Johns Hopkins, 1973).

6. *The Odd Woman* (New York: Alfred A. Knopf, 1974), p. 394.

Not ready to be told. If life as lived truly took the form of stories, and if storytelling were a simple factual reporting or reading off of what had happened, we would usually be willing to tell people just what was going on in our lives. But there are situations in which we find ourselves most reluctant to tell our (immediate) "story." The narrative reluctance we feel at such times is grounded not first of all in a desire to be deceitful — although people do lie at times — but in the *incompleteness* we often encounter in experience. The eventual completeness that allows us to tell other people what has happened and even to enjoy the telling is achieved through further action and/or narrative decisions.

Sometimes a child gets into difficulties at home but does not tell his parents. When they find out anyway, they are likely to ask, "Why in the world didn't you tell us about this?" The answer, which the child may not quite be able to put into words, is that the situation was up in the air, unresolved; it was not the kind of problem that lends itself to a clever or entertaining narrative. What the child had hoped to do was to get it resolved somehow and *then* to tell about it. Had he succeeded, he might even have met with some appreciative laughter in response. The same sort of thing happens with lower-level employees in an organization: they sometimes try to clear up problems on their own instead of reporting them to higher-ups. If they succeed, they may well be told, "You should have reported this when it first came up, but we appreciate the way you took care of it."

In my own experience I can think of occasions when I was in trouble or faced an unusual challenge and then wondered to myself not just what I would — and should — *do,* but also how I would *tell* people about this episode afterward. No "story" seemed to form in my mind as I wrestled with the problem: how I eventually explained the genesis of the challenge, the difficult situation, would depend in part on how it came out, how I managed to deal with it.

We should also ponder what happens when a person contemplates making a major change in his life, such as moving to a faraway city or taking up an altogether different line of work: while the change is in the consideration stage, he may not care to have much social contact with his acquaintances because he does not yet know how to tell people about the possible change. (On the other hand, he will probably seek the advice of some close friends, asking them to keep the matter to themselves.) Once the change has been decided on, a "story" about the decision and its background takes shape and is told to the friends and acquaintances who inquire or come to say goodbye.

Part of the reason for the hesitation in telling the "story" is that people in unresolved situations need to *act;* they have to deal with the difficulty or challenge they face. They do not know how the story will come out because they are not yet sure what they will do or should do, and they want to maintain their freedom of action by keeping their options as open as possible. But there is a bit more to their hesitation, for there are a number of narrative styles or strategies available to them. Those styles are grounded in conceptions of human life and action, and they must choose between them when they get around to telling the story.

A sense of inevitability. Aristotle maintained that a drama, in its presentation of the action, should move beyond mere possibility toward probability to necessity. Hence the action should not be "episodic," and it should not be resolved by a "deus ex machina" (literally: a god from a machine) or something essentially external to the previous action and characters.[7] By the end of the drama there should be a sense of inevitability, which then leaves the audience with the feeling that there has been a genuine ending.[8]

Many people today still think along essentially Aristotelian lines when it comes to the drama that is their own life. Thus, when a decision about a change in their life is to be announced, such people try to maximize the sense of inevitability or necessity. This will make it appear to be a wise, rational decision, one for which there is no need

---

7. Later thinking on drama has expanded the category of "deus ex machina" to include "any forced and improbable device — a telltale birthmark, an unexpected inheritance, the discovery of a lost will or letter — by which a hardpressed author makes shift to resolve his plot" (M. H. Abrams, *A Glossary of Literary Terms,* third edition [New York: Holt, Rinehart and Winston, 1971], p. 39).

8. See Aristotle's *Poetics,* especially chapters 9 and 15. It should be noted that Aristotle meant these observations to apply to tragedy in particular.

to apologize. The factors that seemed to favor the move or decision are highlighted in the story they tell, whereas contrary indications are downplayed. But there are also people whose conception of a good story (paralleling their conception of life and human nature) is that chance or accident or serendipity is uppermost. They will tend to tell the story in such a way that unforeseeable elements, such an unexpected opportunity, determined the outcome.

The choice between these narrative styles transcends mere storytelling, of course; it is to some extent a moral choice. Some people insist on continuity and consistency in their lives; consequently, they overemphasize the elements of coherence that are undoubtedly present. Their tendency is to close off certain life options. Others love surprises and like to see their life story determined at key junctures by things that are totally unexpected.

Suppose a man is giving up a business career and life in the city to take on a small farm in the country. Many of his friends are surprised at the switch. If he values consistency above all, he may well offer a narrative like the following.

> ◘ Perhaps you didn't realize it, but I've always loved life in the country. When I was a boy, I spent a few weeks each summer at my uncle's farm. Now that I'm fairly well off, I want to leave the rat race behind and get back to my real interests. I know there's no money in farming — that's why I didn't become a farmer in the first place — but a farm can be used to offset investment income, of which I have a fair amount by now. I don't plan to do much of the physical work myself — I can hire help for that. I'll mainly be managing the place and enjoying the clean country air.

But suppose his narrative style (and general conception of how life unfolds) leans more toward serendipity. In that case the story he tells people about his decision might sound like this:

> ◘ I'm probably as surprised as anyone that I'll be finished with my business career within two months! My wife always wanted to live out in the country and do some farming, but I wouldn't hear of it. What happened was this: she got an invitation to visit some friends of hers in the country and accepted it without even telling me. I didn't want to go, but it would have been awkward to back out.

> Well, once we were out there for the weekend, I enjoyed myself much more than I thought possible. Now, it turned out that the place next door to the one where we were staying was up for sale. Anyway, before I knew it, I was talking with a realtor about what it would take to buy the place. I remember pinching myself a couple of times and muttering, "This is crazy," but within a matter of three days I found myself making an offer on the place. At that point I was still saying to myself, "I don't have to move out here if I don't want to; I could just hold on to the place for a while as an investment." But when our offer was accepted, there was nothing else to do but start packing. Events sort of made up my mind for me.

It is often difficult, sometimes even impossible, to date in any precise way the real beginning of a story . . . .
–Jorge Semprun[9]

Once upon a time. Children know how stories begin: everything that happened, happened once upon a time, as a children's poem has it. This opening formula transports the child out of his own time into the temporal framework of the story. When adults read (fictional) stories, the opening is generally more problematic — harder to write and a bit arbitrary in readers' eyes. In the case of short stories, the beginning is usually somewhat abrupt: the protagonist is engaged in some action or is thinking about something. In novels there is often a good deal of background sketched in before the action in the narrative present unfolds. One of the problems in launching a large novel in which family connections are a major factor is that a fair amount of "family history," which may seem dry to the reader because it gets covered fairly quickly and includes little narrative detail, must be introduced. Sometimes readers get impatient with all that background, just as readers today often shy away from old-fashioned Russian novels because they contain more characters than one can properly bear in mind. (The fact that the characters have foreign names which

9. *What a Beautiful Sunday!* tr. Alan Sheridan (San Diego, New York and London: Harcourt Brace Jovanovich, 1982), p. 12.

English-speaking readers may confuse with one another probably adds to the difficulty.)

Readers who complain about background and are eager for "the story" to begin might raise the question of relevance: do we really need all those items of family history to follow the story? In the strictest sense of the term "need," the answer is no. But if we ask whether they are of any relevance, whether they contain any potential for shedding light on the story, the answer is yes — we do need them. Historians like to put the events with which they deal in a rather expansive context: in a well-written history book, things just do not happen "once upon a time." Yet the inclusion of a lot of background seems to expand any story to an unmanageable length.

The real limits on a story are dictated not by questions of (potential) relevance but by the capacity of readers or hearers to absorb all the material. How long would it take to read or tell a story if *everything* of possible relevance were included? And how much material can readers or hearers operatively use — or hold in living memory — as background to the key episodes and elements making up the story? Clearly, some sort of selection and reduction must be performed.[10]

The actual practice of many novelists is instructive when it comes to the interplay between (potential) relevance and comprehensibility. Novelists know that their readers are eager to get at the main theme in the story: they want to meet the protagonist and find out what the conflict is that needs to be resolved. Therefore a novelist often begins with a vivid first chapter in which the protagonist faces a crisis that does not get resolved in that chapter; instead the reader is left in suspense. The novelist then hopes we will wonder: who is this person

10. Aristotle therefore placed a limit on the length of a tragedy: ". . . in the plot, a certain length is necessary, and a length which can easily be embraced by the memory." See his *Poetics,* chapter 7, tr. S. H. Butcher, in *Criticism: The Major Texts,* ed. Walter Jackson Bate (New York: Harcourt, Brace & World, 1952), p. 24. The same limits do not apply to novels, which may run on as long as a thousand pages and need not be read in a single evening. Moreover, some fiction comes to us in the form of a trilogy or an even longer series; in many cases, the individual novels are supposed to be readable apart from the series. The writing of historical works is even more open-ended insofar as one could begin reading almost anywhere.

in crisis, and how did he come to such a point in his life? Then the novel may go back in time to the beginning of the story the novelist has chosen to tell and slowly work its way forward to the crisis in Chapter 1. In some cases we do not get back to that crisis until a fairly late point in the novel. There are also novelists who keep advancing the story line introduced in Chapter 1 but slip in frequent flashbacks to earlier episodes in the protagonist's history. Such novelists get in plenty of background, then, but they supply it in small doses. Yet both these types clearly recognize the importance of background, for it is hard to make a clear-cut decision about where a story begins.[11]

It could even be argued that many a story (i.e. novel) begins not with Chapter 1 but with the write-up on the back of the book, or on the inside flap of the dust jacket in the case of a hardcover edition. This introduction of the story officially stems from the publisher but may well have been supplied (at least in its initial draft) by the author. Many readers treat it as part of the text and do not commence reading until they have pondered it; indeed, they may even look back at it from time to time. Well then, if the story in effect begins with such a write-up, in which the conflict central to the story will surely be mentioned (readers want to know what a novel is "about"), couldn't one also argue that for some people the story begins with the account of the book offered by a friend who urged us to read it in the first place? We see again that it is indeed hard to specify just where a story begins!

> *. . . to tell the story as it may appear to some future historian working from the records. . . . This is a story without heroes; and perhaps even without villains.*
>
> –A. J. P. Taylor, commenting on World War II[12]

---

11. Third-rate storytelling tends to ignore background. A narrative weakness in pornographic stories is that there is no background to the action and to the characters' lives: they get right down to business, so to speak, which is presumably what the readers of such works prefer.

12. *The Origins of the Second World War* (London: Hamish Hamilton, 1961), pp. 16, 17. Taylor's approach stirred up quite some controversy, as we can see from the foreword entitled "Second Thoughts," which he included in the edition of his book

Who started it? When two boys get into a fight in a schoolyard, a teacher steps in and asks, "What's going on here?" One of the boys starts to tell the story, only to be interrupted by the other, who insists, "*He* started it!" The issue between them as they argue before the teacher is not just who should get most of the blame but also what would count as a correct or fair narrative presentation of the conflict. If only we had an omniscient narrator on hand, we could ask him for the true account; instead we have to make do with the testimony of the pugilists themselves, perhaps supplemented by some observations from eyewitnesses.

As we look at such a case, we see the necessity of asking, "*Whose* story is this?" The teacher asks both boys for their story. Historians do the same thing when it comes to the outbreak of a war. The two parties to the conflict each have a story, and the stories cannot easily be reconciled: each side is likely to argue that the other side "started it." For example, it might seem clear to almost everyone on the winning side of World War II that Hitler and company had initiated the hostilities, but even in this case it has been argued that harsh peace terms imposed on Germany at the end of the first world war had a great deal to do with the outbreak of the second great war. So what is the story of Hitler's war? Is it a story that begins with some thugs who happened to come to power in Germany and then began making trouble for their neighbors in hopes of getting a major war going? Or is it a story that begins with unwise and unrealistic statesmanship at the end of World War I?

The case of World War II reminds us to what an extent history is written by *victors:* the very genres in which stories are told seem to favor victors. On the other hand, the aftermath of the war has also given rise to a new appreciation for *victims'* stories. Yet the stories told by victims often seem flawed or weak from a formal narrative standpoint. Consider the following sad victim's story, which comes from a non-omniscient narrator whom we must imagine telling his tale in the life to come:

---

published by Penguin Books in 1963.

> After they moved us all to the ghetto and sealed it off with a wall, they left us alone for quite a while. Then one morning, without any warning, we heard them come into the ghetto. There was shouting, cursing, and some gunfire. They rounded up our entire block and marched us off to the train station. They put us on a filthy train fit only to carry animals, and we rolled off to the east, for about two days. Some people thought we were being resettled in a rural area; others wailed that we would be shot. We finally got to a camp where we were told we would have to strip and take a shower: we wouldn't even be allowed to keep our clothes, which by then were filthy. We were jammed into a big shower room and then a horrible gas was pumped in. That's the last thing I remember . . . .

Even though the events depicted here are frightful, the story is somewhat flat and uninspiring. Such accounts have often led people to ask: but couldn't they have fought back? And so, when we read or hear stories about victims, we prefer the ones in which the victims *did* resist. One reason why Jean-François Steiner's book *Treblinka* is such a stirring story is that it presents us with victims who opposed their oppressors and even carried out a successful rebellion. However much we may sympathize with victims, we yearn to find a victor to identify with when we hear a story.

The moral of the story. The notion that the appropriate ending for a story is a moral-practical lesson is an old one: we are familiar with it especially from fables of the sort that are usually attributed to Aesop, in which the lesson for the story is spelled out in a single line, such as "Familiarity breeds contempt" (The Fox and the Lion), or "One man's meat is another man's poison" (The Ass and the Grasshopper), or "A bird in the hand is worth two in the bush" (The Angler and the Little Fish), or "Honesty is the best policy" (Mercury and the Woodman).[13] Implicit lessons are easily drawn from many

13. See *Fables: Aesop and Others* (London: J. M. Dent & Sons, 1913), pp. 17, 25, 37, 45. These fables are also published in editions that omit the morals: there is even an edition called *Aesop Without Morals* (New York and London: Thomas Yoseloff, 1961). Its editor, Lloyd W. Daly, complains that the stated morals are "little more than an insult to our intelligence" (p. 12).

other stories and novels as well, even though English teachers on higher levels usually claim to oppose "didacticism."[14]

A great many first-person narratives, as we encounter them in daily life, also contain a "moral," but not one that is of the universal-lesson variety. Very often the point of a (presumably true) story that someone tells us about himself is *self-justification.* A man may say to his friend, "I just heard something very strange about you, something I have trouble believing, and I want to ask you directly if it's true." The response might be, "What you heard is true as far as it goes, but please take the time to listen to my story." Once the (whole) story has been told, the single item that had been passed on to the man making the inquiry stands in a fuller context; the man telling the story has succeeded in justifying himself, or at least in minimizing the extent to which the events place him in an unfavorable light.

The need to draw a moral conclusion about one's actions is part of the reason why people are often reluctant to tell about events they are involved in while they are still unfolding. The reluctant person has a sense that the story is not yet rounded off and therefore is not properly "tellable." It will be rounded off, as far as he is concerned, when (i) he resolves it by taking some appropriate action, or (ii) he thinks it through sufficiently to make up his mind what it is that he has learned from it and is willing to say about it,[15] or (iii) some combination of (i) and (ii).

Suppose a child has broken something of value in the home. He may keep his deed from his parents while attempting to fix the broken item. Perhaps he knows from the start that his attempt will not succeed, but he wants to add the story of the repair attempt to the original story. The fuller story will place him in a somewhat better light: he indeed did wrong when he broke something he should never

---

14. See the Glossary.

15. Not in every situation is learning the acquisition of something *new*. Sometimes a story begins with the declaration: "Today I did something really stupid" and ends with: "That just goes to show that you can never . . . ." The content of the lesson reinforces something the narrator already knew but failed to act on in this instance. By putting the moral of the story up front, he prevents others from commenting on his stupidity — he beats them to the punch, so to speak.

have touched in the first place, but he made a valiant effort to repair the damage. In all likelihood, the attempt to repair the damage will result in a more lenient response on the part of his parents.

Storytelling contains a greater moral component than many people realize. Moreover, drawing or pointing toward a lesson is one of the standard ways to wrap up a story. Yet I would stop short of affirming that all literature and storytelling must be regarded as a species of moral discourse. Storytelling, after all, is something human beings do, and not all humans are alike. Some are much more committed to the legitimacy and centrality of moral considerations in human life than others are. Thus there are many people, both writers and tellers of (presumably true) stories, for whom the moral dimension is indeed the key to the narrative, but there are also people for whom morality in a traditional sense does not enter into a story. Perhaps we could speak of their stories as having an anti-moral; they are intended to bolster and underscore the nihilism in which so many people live. Such people do not produce stories that satisfy or greatly interest *me;* in the final analysis, their stories seem pointless. Yet I must acknowledge that my analysis does not shed much light on what they are up to when they tell stories.

Making peace. When the end of the story comes, it is time for us to go back into the "real" world — unless, of course, we choose to hear or read another story first. Does the way we end our stories contribute to life in the world? It probably does in various ways: one of the most important, surely, is the potential which stories have to make and strengthen *peace* between human beings. Yet this dimension of storytelling is often lost from view.

The problem, again, is that the stories in circulation tend to be told by the victors. The vanquished may well have perished; and if they did survive, they may not dare speak up; and if they do speak up, it may be in a desperate tone that breeds distrust.[16] Surely, one of

---

16. In many terrorist incidents, the people committing the terrorist acts demand that the world listen to their grievances: they will not be satisfied until a certain story (their story, as members of a vanquished and oppressed group) is made known

the lessons to be learned regarding storytelling is that victors need to be gracious in telling about the battle once it is done. It would be a mistake to humiliate the vanquished once the conflict is over: we must all live in this world together as neighbors.

The question of the Holocaust and of the atrocities committed against all sorts of human beings before and during the World War II provides us with a useful example of peacemaking through storytelling. The truth of the matter is that those horrors were unleashed by the Nazi regime which had taken over in Germany and which managed to enlist a great many Germans in its ghastly program of brutality and degradation. Some people therefore feel justified in saying, "The Germans tried to wipe out the Jews, and they committed atrocities against many other people as well during World War II." A truer and fairer statement, and one that would contribute to peace and harmony, would be this: "People of all the countries that were occupied by the German army participated in Holocaust activities and other atrocities along with Germans, and people of all those countries (and Germany as well) offered heroic resistance to such crimes against humanity: many of them paid for their resistance with their lives." The latter statement leaves more room for the Germans of postwar Europe (remember that today the German nation is made up largely of people who were born *after* the Nazi era) to regard themselves as members of a civilized human community that unequivocally condemns the thinking and actions of the Nazis.

Peacemaking is also a factor in the stories that emerge from marital disputes. A husband and wife may have a nasty, noisy fight one day, and finally reach an uneasy truce after a couple of hours of shouting and screaming. They have not resolved all their differences, and they realize that there will be future arguments. Nonetheless, they must return to their everyday routine, and so they find themselves asking: "What shall we tell the kids?" Their children, of course, could not help but be aware that a fight was underway. What is then needed is an account which does not deny that there was conflict (for such a denial would sow distrust between the children and the parents) but

---

to the world over radio and television networks.

which does embody a mutually agreed upon version of the *reason* for the conflict, e.g. disagreements about money. The experience gained in coming up with such an account is excellent preparation for conflict resolution in other sectors of human life.

The Christian emphasis on loving one's neighbor (the second of the "two great commandments") can be understood in part as a commitment to the kind of storytelling that respects the need for peace. The second great commandment is supposed to embody the heart of a number of commandments, including the injunction not to bear false witness against our neighbor. In both the heat and the aftermath of battle, this commandment is hard to observe scrupulously. (Husbands and wives who have been fighting often cannot help belittling and running down their spouse, even in the presence of their own children.) And so we need to be reminded that the stories we eventually tell must help to establish peace on earth.

> *. . . we will have ceased to rewrite our autobiographies only when we will have ceased to be (we don't merely add on to them as the years go by) . . . .*
>
> –G. B. Madison[17]

Inconclusive endings. Many stories children read contain tidy endings: after the villain is dispatched, we are told that the young lady in distress married the handsome prince who rescued her, "and they lived happily ever after." Autobiographies are not like that: when we come to the end of a particularly good one, we cannot help but wonder what comes next. If the author has already died, we wonder what his last years or days were like. (He has not told us in the book.) Or if he is still alive, we may hope that more of his story is forthcoming. And if we read the autobiography of a reasonably young person, we are aware that there may be later "chapters" in his life that will lead to substantive revisions in the story he tells about himself. Many of us have felt the same need when drawing up a short account of our own

17. "The Hermeneutics of (Inter)Subjectivity, or: The Mind-Body Problem Deconstructed," in *The Hermeneutics of Postmodernity: Figures and Themes* (Bloomington and Indianapolis: Indiana University Press, 1988), p. 165.

life on a job application: if we want the job badly enough, we will wind up sounding as though our entire life has been a preparation for something just like the opening for which we are applying. And so the stories of real lives seem more inconclusive than the tidy stories children love to read.

There are some novels that come to an exciting, dramatic conclusion that we remember for a long time afterward, but a great many seem to trail away. Occasionally I am asked how a novel which I read some time ago came out and I find that I cannot remember, although I can say a fair amount about the characters and the central conflict. The conclusion, it seems, was not memorable. And sometimes I have the feeling, when I am in the last twenty to fifty pages of a novel, that the story is basically over, and that I would not miss anything if I were to discontinue at that point. There are also cases in which I enjoy a novel through the first half or two thirds but feel no need to pursue the story to the end. And since I do not read novels professionally,[18] I allow myself to discontinue. After all, there are many more interesting novels to read than any one person could ever get through in a single lifetime.

What we learn from a careful consideration of how people actually read is that the ending is a weak feature in many a novel. Some novelists seem to practice a version of the "moral" approach to stories: the protagonist must be seen to have grown in some way, to have come to a new awareness or appreciation, or perhaps to get his "comeuppance," if that's what he deserves. Once this has been signaled or adumbrated, the narrator can fade away, and one has the sense of an ending. But such an ending requires that the novel itself be somewhat upbeat. In many pessimistic or darker novels, such a conclusion cannot be offered, and as a result the ending seems weaker, inconclusive. But *life* is often said to be inconclusive as well — this, presumably, is the point of such a novel. The weakness in ending is often paralleled by other weaknesses in plot structure.

---

18. I am not an English professor, and so I am under no obligation to be widely knowledgeable when it comes to novels.

> *. . . I know the ashy taste of not living up to some part of your dream. I even know the necessity of making constant adjustments to your life story, so you can go on living in it.*
>
> –Gail Godwin[19]

The need to entertain. In our commitment to truth we sometimes instruct people to be sober and factual in their reporting or testimony: "Don't varnish the facts. Just tell us exactly what happened — all of it, in order." If people comply with this instruction, however, the narrative they produce is not likely to be very interesting. They then bump up against a fact of life we must all reckon with, namely, that it's not easy to command an audience, and even harder to hold one. If we want people to *listen* to us, we should be entertaining talkers, *raconteurs.* And it's no simple feat to win such a reputation, for when it comes to storytelling, we face stiff competition.

This point has been brought home to me personally on occasions when I listened to some unfortunate person's tale of woe and found myself drifting into aesthetic criticism of it, thinking, "I've heard other stories like this one, but they were much more interesting." Indeed, classic literature presents us with many tales of woe, but the stories are told in an engaging way that holds our attention — otherwise people wouldn't go on republishing and rereading them. Ordinary supplicants who need our help have to compete with professional storytellers, then, and a few of them do become quite skilled in the narrative arts themselves — to the point that some people may consider them charlatans. Is there a happy medium between honesty and narrative embellishment?

I don't dare say there is. Rather, it occurs to me that the willingness to listen to someone's tale of woe, however ineptly it may be told, is largely a measure of one's commitment to that person, and perhaps of one's professionalism. There are, after all, people who make it their day-to-day business to talk with those who are having difficulties and to hear them out, however long it takes. Some of them are paid by the

---

19. *The Finishing School* (New York: Viking Penguin, 1984), p. 322.

storyteller; others are on the payrolls of hospitals, social service agencies, churches, or charitable organizations.

The existence of psychotherapy (which requires its practitioners to listen to seemingly endless narratives in which people in distress allow their story to dribble out ineptly) should serve as a warning to us: the artful construction which many people are able to work into their narratives has the potential to mislead us. A psychotherapist knows that he will not make much progress with a patient who is highly prepared for his encounter and has the story all ready in a neatly written and edited form (comparable to what we sometimes put on job applications). Much of what happens in psychotherapy is an effort to get beyond the carefully prepared and rehearsed answer or story. The raw material does not come out easily; that's why psychotherapy can often be such a frustrating and drawn-out affair.

The contrast between psychotherapy and the novel should serve as a reminder to us not to polish our stories too much. Nowadays, one's oral performance in a job interview needs to be a careful blend of artful presentation, on the one hand, with inconclusive hesitance, on the other: it is getting harder and harder to come across as sincere.

Because we do go through life making adjustments to our story on the basis of new realizations about ourselves, the professional polish we find in storytellers, who can make even trivial incidents enchanting, is not something we should all strive for. Yet we should note that ministers need to be accomplished storytellers if they are to be effective as preachers. It is healthier to look for the truth concerning oneself in some pre-narratized material, relying on the friendly and sympathetic assistance of a therapist or counselor who is committed to our well-being, than to "know it all" right from the outset. There is indeed some truth in our narratives about ourselves, but it should not be confused with polish.

Chapter 6

Four Constitutive Principles

History is a language-net cast backward.
–George Steiner[1]

Understanding in retrospect. The four principles which I will explore as "constitutive" for narrative consciousness and historical thought are anachronism, anthropomorphism, polarization, and integration. I will consider why — and in what sense — each of these is indispensable for gaining and maintaining a sense of the past as the background and context for our lives.

Perhaps the necessity of anachronism can best be grasped by way of the notion of a narrative thread. There are situations in which human beings find that they have only a vague, indistinct, blurry sense of the past, of what has gone before. Their confused state of mind may be due to illness or intoxication or some other abnormal condition; but whatever its source, we know that it keeps them from functioning as effectively as they might. They seem disoriented and are aware of needing a clearer sense of the immediate past, in particular. They may ask others about the part of their past that seems blurred and out of focus.

Yet the ideal condition is not *total* recall, for such is not possible for finite minds. We can never get beyond a *partial* grasp and retention of what has come before. Whatever degree of retention we manage to attain, we can be sure that it will include temporal gaps: strict sequence is not the thread that enables us to make sense of the whole. What I have called the narrative thread probably has many

---

1. *Extraterritorial* (New York: Atheneum, 1981), p. 62.

components, including some that will not be explored in this essay, but anachronism (violation of temporal sequence, whereby something later shapes something earlier) is one of the most important of them.

A remarkable example of the functioning of anachronism in ordinary consciousness and memory occurred for me on the very day this section was written. I was walking briskly down the hall at Redeemer College when I saw someone who looked familiar. I walked up to him and asked, "Aren't you Harry's brother?" Now, Harry Zekveld is a student of mine, and I was aware that his older brother Tony would also be a student at Redeemer this year. It happens that Tony had earlier studied at Redeemer during the 1983–84 academic year, and although he did not take one of my courses, I did get to know him and so I remembered him — in roughly the way we remember people we have not seen for years; that is to say, I could call up a mental image of him. But when I greeted him in the hall, what I "saw" first was his brother Harry's face, which I know so well, since Harry and I have had a great deal of contact. Now, Harry had come to Redeemer to study *after* Tony left, and so my knowledge of Harry could have no effect on my perception of Tony. Yet today I could not help but view Tony within the framework of the Zekveld countenance as I know it through Harry: it was Harry's face I spotted, so to speak, and then saw that it wasn't really Harry after all, but someone who looked like him. After a brief conversation with Tony, I moved on and tried to bring back my memory image of Tony, which earlier had been uninfluenced by my contact with Harry. But now the features Tony has in common with Harry stood out in memory. Those features were also there back in 1983–84, of course, but at that point I did not look at Tony in the light of Harry because I did not know Harry yet; now I cannot help but do so.

What happened in the Zekveld case is unavoidable. The conclusion to be drawn from it is that the necessity of anachronism is first of all psychological: we are so constituted as to be influenced by later experience and knowledge in our retention of earlier experience. But the additional consideration I wish to explore is that our concentration on similarities helps render human experience in its sheer abundance somewhat more memorable; it gives us a thread, as it were, for linking the many items in memory and consciousness in such a way that we are able to have a sense of the whole. Hence it is no accident that we so often notice and comment on physical and behavioral similarities between members of a family as we come to know them: we need those similarities to keep some order in the store of knowledge we call memory.

The alternative to allowing anachronism a positive role in the functioning of our continuing consciousness of the past is to strive for the (sterile) purity of a chronicle. By a chronicle I mean a written record which is composed at the time events unfold and which is therefore not open to revision and recasting in the light of what comes later: it is only added to. Naturally, a chronicle cannot help but be influenced by the past, that is, by the categories and concepts developed before the events recorded in the chronicle took place, but it is left untouched by later events and realizations. To go back to my Zekveld example, a snapshot of Tony in 1983 cannot be changed by my subsequent familiarity with Harry, even though my *viewing* of that snapshot today probably would be. The snapshot itself is the chronicle, so to speak; it is a pure particular, unaffected by subsequent events and developments.

The fact that chronicles play a relatively insignificant role in our relationship to the past demonstrates the indispensability and value of anachronism. A strictly "chroniclized" approach to the awareness of the past will not serve the purposes of "life" (Nietzsche's emphasis). In the end, a massive, lengthy chronicle will turn out to be indigestible and will border on incoherence.

But here I need to pause again for a moment to acknowledge the great fear people rightly raise in connection with the acceptance of anachronism, namely, that it seems to leave the door open to the cynical falsification of the past. That there is a danger here cannot be denied; the only response I can offer is to say that there are some possible safeguards, some factors that are able to contribute to truthfulness in our narratives and memories. I intend to explore those factors in Chapters 7–9.

The hermeneutical circle.[2] The defense of anachronism I have offered here needs to be placed in a larger framework if it is to be properly appreciated. The problem of the hermeneutical circle is familiar to many people because hermeneutical considerations come up in various sectors of scholarship, including the study of the Bible; indeed, for some people the term "hermeneutics" is virtually equivalent to Bible study. Now, the way the Bible has been interpreted by orthodox Protestantism is instructive for us, for what we find when we look closely is that the phenomenon of the hermeneutical circle is nicely exemplified there.

The problem can also be stated as similar to the old chicken-and-egg question. (Which comes first — the chicken, or the egg?) Which should we regard as prior — the meaning of the whole, or the meanings of the individual parts? On the one hand we are inclined to say that the parts are first understood separately; then follows the meaning of the whole as a kind of aggregate. On the other hand, a little reflection on how we actually comprehend sentences shows that no separate and prior assignment of meaning to individual words takes place: what a word means depends on the (meaning of the) sentence as a whole.[3] There is an interaction, a movement back and forth, between the meaning of the sentence as a whole and the meanings of the words that compose it.

In Biblical theology we recognize this point when we speak of "progressive revelation." To understand any part of the Bible, including the earliest chapters of the Old Testament, in a truly Christian way,[4] we must bear the meaning of the entire Bible in mind. Yet the

---

2. See the Glossary.

3. For speakers of English, this point becomes painfully apparent if they study German, a language in which many sentences locate a key element of the main verb (surely an important word for ascertaining meaning) at the very end. People complain about this phenomenon: their complaint illustrates how much a sense of the whole is needed (and provisionally assumed) when we try to decipher the individual parts, i.e. the words.

4. It is Christian in that it emphasizes the centrality of *Christ* for the Biblical message. Jewish theology, with its refusal to recognize Jesus of Nazareth as the Messiah of which the Old Testament Scriptures speak, will of course come up with

meaning of the whole Bible is not to be ascertained apart from a careful study of the meanings of the various books and parts that make up the Bible. Still, the sense of the whole enjoys a kind of priority, so that we could say that our understanding of the Bible (especially in Reformed thought) is shamelessly anachronistic.

We should not suppose that this openness to anachronism is merely a characteristic of theologians who think and write about the Bible: it is to be found in the text of the Bible itself, as we see from the way it uses the name "Yahweh" or "the LORD."[5] In the famous story of the burning bush, we read that God *first* revealed himself as "Yahweh" or "the LORD" to *Moses* (Exodus 3:13ff; see also 6:3). Yet we read in earlier passages that *Yahweh* (the LORD) spoke to *Noah* (Genesis 7:1), to *Abraham* (12:1), and to *Jacob* (28:12ff). In the Jacob instance God is even quoted as saying, "I am the LORD . . ." (vs. 13). Indeed he is and was, but we also know from the book of Exodus that he had not yet revealed himself as such. Nevertheless, this telling anachronism is permitted in the inspired record of his dealings with his people. Even more significant is the fact that we find Jacob saying, "Surely the LORD is in this place . . ." (vs. 16). In substance Jacob presumably did say this, but he must have used different words! The later revelations sanction an anachronistic retelling of this encounter between Jacob and his God.[6]

---

a different slant on many passages and elements of the Old Testament.

5. "Jehovah" in some translations. Biblical theologians in our time generally stay away from the name "Jehovah" and prefer "Yahweh." In many translations "the LORD" is used, but there is of course no oral distinction between "LORD" and "Lord."

6. The awareness that Biblical language (like our everyday discourse) is awash with anachronisms should also be brought to bear on our understanding of more ordinary texts, such as the one that tells us that Eve "conceived and bore Cain" (Genesis 4:1). In what sense was he "Cain" at the moment of his conception? Did Eve set out to "conceive Cain"? In the New Testament we find Stephen saying that "the patriarchs" sold Joseph into slavery in Egypt (Acts 7:9). Were they "the patriarchs" at the time they did this — or were they a band of brothers jealous of the one favored by their father? When we ponder such questions, we see again that the Bible is *God's* book — the record of *his* dealings, which are beyond time and anachronism.

. . . the perception or creation of analogies and likenesses is the central function of all understanding . . . .
–G. B. Madison[7]

Understanding others. An agent that is completely alien will not be easily understood by us.[8] (This is part of the reason for our fascination with stories about creatures from outer space, with whom we may have virtually nothing in common physically and mentally, and surely nothing historically and culturally.) Yet human experience testifies that an understanding of sorts is eventually achieved when intelligent beings spend time together. The reason this is possible is that we seek to understand the alien (or anyone, for that matter) on the basis of — and in terms of — *our own experience.* To the extent that we can find or establish points of likeness between the alien and ourselves, we have something to work with in our attempt to comprehend him or it.

This approach to understanding can also be demonstrated in our efforts to understand members of the opposite sex. For all human beings, difficulties will sometimes arise in this area, such that we might throw up our hands and say, "I just don't understand women" or "I don't understand men." Yet there is considerable basis for understanding, since men and women are alike in many respects, both physical and mental. The points of difference need to be approached via imagination. I can use my imagination to get some sense of what it is like to be a female, thereby enlarging my own experience as the basis for my acts of understanding, but I always draw on my own bodiliness. We can also use imagination to develop a sense of what it would be like to be without arms or legs, which is the plight of some human beings. There is, then, no reason for despair when it comes to understanding those who are different from ourselves in some important respect; if only we are willing to work at it, our culture will provide us

7. *Understanding: A Phenomenological-Pragmatic Analysis* (Westport: Greenwood Press, 1982), p. 153.

8. I use the term "agent" here to mean something that acts. To use "person" would be to assume some sort of commonness between the "agent" and ourselves.

with some resources we can draw on to enlarge our experience.

In our understanding of God, who, surely, is fundamentally different from us,[9] the doctrine of the image of God plays an important role. The assurance that we have been created in his image and likeness gives us reason to expect that it is possible for us to fathom his ways and doings to some extent and to share in them in a small way. And this likeness does not only apply to theology or *thinking* about God: we are to be *doers* of the Word (James 1:22), ministers of the love of God, channels of his grace. "Be merciful, even as your Father is merciful" (Luke 6:36). God forgives, and we are to forgive as well. "How is all of this possible?" one might ask. The answer is that we can do these things because we are *like God* in important respects.

Now, this process of using my own experience, including my physical and mental make-up, as a springboard for understanding those who are different from me, is properly called anthropomorphism. This term is rooted in a pair of Greek words meaning "man" and "form" respectively. In general, "anthropomorphism" means likening something, such as an animal, to a human being and proceeding to understand it in human terms. In the analysis I am presenting here, I would stress that we do not understand things in *generally* human terms but rather in more specific terms; that is to say, we understand in terms of *our own experience.* In other words, the "anthropos" (man, or person) is *myself.* My own experience, including what I have been able to imagine, is the basis for my acts of understanding.

Anthropomorphism is a constitutive principle of narrative consciousness and historical thought. If we did not engage in this cognitive strategy, much of present human experience (to say nothing of our approach to the remnants and records of the past) would confront us as forever foreign and incomprehensible. We would look into the mirror of history and see only ourselves — all the rest of it would be a confused blur. And we would not fare much better when looking at social relations in the present. G. B. Madison maintains: "Chaos, the absence of determinations, cannot be perceived and repels the mind."

9. "Alien" is not a term one would use in connection with God, but Karl Barth conceives of him as "wholly other."

The human being is a maker of meaning: he is driven to understand. And understanding, which Madison does not regard as fundamentally different from perception, involves apprehending one thing in terms (already familiar) of another, that is, in terms of our own prior experience. "To understand," he writes "is to interpret; to perceive something is to see it as some thing."[10]

When we recognize the role of anthropomorphism in historical thought and experience, we can see why historicism[11] has cut such a wide swath through twentieth-century intellectual life. If everything we discover, learn and remember about the past must somehow relate to my own experience, perhaps there is only historical truth *for me,* historical truth that *relates to me.* In that case, any claims to objectivity with regard to the past would have to be discarded. (A response to the challenge of historicism will be presented in Chapter 9.)

Personifications. When we seek to understand others in their role in history, their *actions* are the object of our special interest, for it is their actions that bring them into relation with people and forces and ultimately with ourselves in the present. To understand others on the basis of analogy with our own experience is thus to think of them as acting in roughly the same way that we ourselves act.

That we do conceive of the other in some such manner is evident from the way we deal with animals when we make them protagonists in stories: we personify and humanize them. We draw them in a quasi-human form and assign them human roles, as in Richard Scarry books intended for young children. And in stories for slightly older readers, we endow them with thought processes (and therefore language) much like our own: I think, for example, of Jack London's novel *White Fang* (published in 1906), which is about a dog in the far north. The protagonist of this novel enjoys an inner life of thought that renders its actions explicable and secures a narrative thread that runs throughout. It is also possible to talk about animals and their actions in purely physical terms (in which case we should call them movements), just as

10. *Understanding,* p. 73.
11. See the Glossary.

we might talk in strictly physical terms about what round balls do on a pool table when they bump into one another; but in that case we would not have a *story*.

We face a bigger challenge when we must deal with corporate agencies (companies, governments, armies, and so forth). To be able to talk about them as *acting*, as having motives, plans and intentions, we must again personify them. Hayden White makes much of "tropes" or figures of speech in his analysis of historical discourse.[12] One of the tropes discusses is especially pertinent here — metonymy. In this figure of speech we see the process of reduction underway: some part of a larger whole, perhaps the key part, is spoken of simply as though it were the whole.[13] We engage in such reduction frequently in everyday language, and historical narratives are shot through with metonymy as well. How does metonymy help us talk about action in history?

Its main benefit is that it renders the historical arena somewhat more comprehensible than it would otherwise be, and this is its ultimate justification as a form of anthropomorphism. Suppose there are significant geopolitical developments underway in the Middle East. People ask: what will the United States do? How will the United States respond? Now, the name "United States" covers an enormous amount of territory (literally, and also in terms of the people it applies to, not all of whom live within U.S. borders). The name also refers to the agencies, departments and offices that together make up the government of the United States. It is hard to think clearly and

---

12. See *Metahistory* (Chapter 5, Note 5), especially pp. 1–42; see also the Introduction to *Tropics of Discourse* (Chapter 5, Note 5), pp. 1–25. It should be noted that White's use of this term does not correspond fully to definitions offered in standard handbooks of literature. There are various ways in which the figures of speech we use can be classified and defined. In drawing up his list of four tropes that are especially relevant to historical representation, White appeals to the work of Kenneth Burke. For Burke's discussion of these "four master tropes," see *A Grammar of Motives* (Berkeley and Los Angeles: University of California Press, 1969), pp. 503–17.

13. Metonymy comes close to synecdoche in terms of how it functions: indeed, sometimes the latter is regarded as a variation on the former. White's standard example of metonymy is "fifty sail" as standing for "fifty ships."

concretely while using such an expansive term, and so people may well rephrase the question by asking: What will *Washington* do in response? Washington is of course the capital city of the United States and the place where a good part of the government is housed. Somehow it is easier to think of "Washington" as trying to respond to new developments overseas. But this version, too, may be too general, for there is much more to Washington than just government activity, and so we find ourselves asking: What will *the White House* do? The White House is a building in which the President of the United States has his office (and his residence); many members of his staff have offices there as well. It is the site of meetings attended by the President, and important decisions are often made there. Yet it is not the whole of the government — far from it. The State Department and the Defense Department both have impressive facilities of their own separate from the White House. Thus, although it is more concrete to talk about the White House and what it might do than about "Washington," it is not very accurate in terms of who wields power and has an important say in what the United States does in foreign affairs. And so we may find ourselves going one step further and asking: What will *President Bush* do? Now the picture becomes quite concrete. Mr. Bush, after all, is a human being, and we can well imagine him functioning in roughly the same way that we function ourselves. In moving from "What will the United States do?" to "What will President Bush do?" we have used a figure of speech — metonymy.

While it may be helpful to use such a figure of speech and substitute "President Bush" for "the United States," it can also be misleading and inaccurate. There is a sense, of course, in which everything the government of the United States does is the responsibility of the President, so that we might speak of him as fighting pollution in Pittsburgh, even if he never sets foot in Pittsburgh while serving as President; yet on an operative level he does not "do it all," nor is he aware of all that goes on. Suppose a boat is making its way under cover of night to a remote berth somewhere on the coast of Florida. It is conceivable that drugs are being smuggled into the country on this boat, or that human beings are entering the country illegally. What will the United States do? In this case President Bush does nothing and presumably knows nothing. Instead, agents of the United States government on the local scene deal with the matter in accordance with the laws and policies in effect. Concretely speaking, the Coast Guard intercepts the boat and finds out who and what is on board. It does not advance our understanding of how government works to insist that in some remote sense President Bush (who is asleep in the White House in Washington) has intercepted the mysterious boat.

If we consider cases in which *illegal* activity is undertaken by the government, the metonymic understanding gets stretched even further. When Richard Nixon was President and members of his staff engaged in illegal activities, it became an interesting and important matter to assess the extent to which the President himself, either actively or by omission, could be said to be a participant in those illegal acts. Similar questions have been raised about the Holocaust: to what extent did prominent members of the Nazi regime have specific knowledge of what was going on? The law has long recognized that one need not have participated in an action in the ordinary, common-sense meaning of "participate" to be guilty of complicity and to be liable before the courts. Parents are usually liable for damage caused by their children.

While the form of anthropomorphism known as personification brings benefits in terms of rendering certain types of action more thinkable than they would otherwise be, it is far from unproblematic. And the difficulties it involves grow still greater when we turn to joint action and interaction. In the executive domain, responsibility is exercised by the so-called executive branch of government, but when it comes to making laws and approving taxes and expenditures, the Congress of the United States also enters the picture. With its hundreds of members, of course, the Congress is a more difficult entity to understand than President Bush, who is an individual human being and can be said to put his pants on one leg at a time. Because the Congress and the President share responsibility in various areas that have to do with spending public funds and making new laws, the question "What will the United States do about such-and-such?" must sometimes be interpreted as the question "What will *Congress and the President* do about such-and-such?" And since there may be considerable divergence of political opinion within the Congress and also between Congress and the President, the action taken will probably involve a political tug-of-war. In this tug-of-war, *anticipation* will be a factor: as the one side speaks or makes a proposal, it will have to anticipate the reaction or response of the other side. When a decision is finally made, it will be hard to trace the process whereby it was reached: such decisions are much more mysterious and unfathomable than decisions made by individual human beings. An individual can give reasons and point to factors that influenced him considerably. But if one asks a member of the Congress why such-and-such a decision was finally made on a budget issue, he can at best review the process of negotiation and political in-fighting that led to the decision. The Congress as a whole does not have motives and intentions in the same way that an individual human being does.[14]

> *No constructors, no deconstructors. No norms, no perversions. . . . Everybody needs everybody else.*
> –Richard Rorty[15]

Positing a non-ego. A theme in philosophy which non-philosophers usually have difficulty grasping is subject and object. Not only are the "answers" hard to understand: one does not quickly see what the question is either. Most students of philosophy realize that the question of "subject and object" has something to do with the nature of knowledge; some of them (especially if they have studied nineteenth-century German philosophy) recognize that this theme also bears on the nature of human action.

The German thinker J. G. Fichte (1762–1814), who is not very

14. Since I am a Canadian, I might have been expected to use Canadian political examples. I have chosen to talk about the United States Congress by way of example because the Canadian (and British) parliamentary system, with its closer relationship between executive and legislative functions of government, is more unified and therefore not as well suited to illustrating the philosophical point I am trying to get across.

15. "Philosophy as a Kind of Writing: An Essay on Derrida," in *Consequences of Pragmatism* (Minneapolis: University of Minnesota Press, 1982), p. 108.

well-known in English-speaking countries, even though he is part of the philosophical path leading from Kant to Hegel, spoke of the necessity (ultimately moral in nature) of positing a "non-ego." Human existence is of such a nature that the ego (the human person as knower and doer) needs something other than himself (a non-ego) to make it possible for him to know and to do.[16] This analysis represents another version of the subject/object scheme: the subject needs a "Gegenstand" (object)[17] in order to be able to be subject.

What Fichte contributes to our discussion is a notion I would prefer to call *polarization.* I mention it here because it is a constitutive principle of narrative consciousness and historical thought. One of the inescapable givens of the human situation is that we need to polarize, to posit an object (often interpreted as an opponent, a "tegenstander"[18]). The opposition between subject and object, ego and non-ego, is sometimes characterized as "dialectical" in nature. Fichte is therefore called a dialectical philosopher, and so is Hegel, who is better known than Fichte for his commitment to the notion of polarization. It was Hegel's emphasis that oppositions, as they spring up in consciousness and in the sphere of action, are usually muted and overcome in time, even though a new bifurcation arises to replace them: "theses" are constantly being confronted by "antitheses."

The notion of polarization as essential to narrative can also be expressed by means of the term "conflict." We might then say that wherever we have a genuine story, we have conflict. Many stories are about heroic adventures and fierce battles, of course, but what about

---

16. According to Fichte the non-ego is necessary if I am to be able to perform my duty. In the words of Frederick Copleston: "The natural world, the sphere of the non-ego, can thus be regarded as the material or instrument for the fulfilment of our duty . . ." (*A History of Philosophy,* Vol. VII [Westminster, Maryland: Newman Press, 1965], p. 65).

17. This German word for object is often quoted in philosophical discussions because it nicely brings out the idea in this term: the object is that which stands opposite or facing (*gegen*) the subject, ready to be acted on, or perhaps to offer resistance. The same idea can be derived from the Latin etymology of the word "object."

18. This Dutch word for "opponent" shares etymological roots with the German *Gegenstand.*

the stories that have "love" as their theme? A careful look at love stories will show that conflict of some sort (usually not physical) between the lovers is what makes the story go. We also get the impression that relationships are not stable: where there may appear to be peace and harmony, discord quickly springs up (a point Hegel would appreciate). When we speak of polarization as a constitutive principle of narrative consciousness and historical thought, we are recognizing that without it we would have no story, no possibility for narration and for the kind of memory it makes possible.

> *I will sing to the LORD, for he has triumphed gloriously; the horse and his rider he has thrown into the sea. . . . The LORD is a man of war.*
> Exodus 15:1, 3

Living with conflict. In our time many people fear conflict and do their utmost to minimize it. Some even interpret Christianity as rejecting any meaningful role for conflict in human life: we are always to "turn the other cheek" (see Matthew 5:39). Yet such a picture does not fit the conduct of Christians throughout the ages, nor does it fit their God, who reveals himself as a "man of war." Christians in the Middle Ages were fond of conflict (think of the Crusades) and enjoyed verbal battle and polemics. Indeed, the language in wide use among educated people in the Middle Ages (i.e. Latin) was dominated by a pugilistic, polemical spirit: it has even been argued that it was not a language suitable for girls and women, in whom a gentler spirit is presumably to be found.[19]

Conflict is also an everyday theme in the twentieth century: the

---

19. Walter Ong suggests that "learned Latin" (perhaps we should speak of the *spirit* of the medieval Latin language) could not be taught to females. It appears that the introduction of females in large numbers into the world of higher education sounded the death-knell for Latin as a subject in schools. See "Latin Language Study as a Renaissance Puberty Rite," in *Rhetoric, Romance and Technology* (Chapter 2, Note 27), pp. 113–41; see also "Latin and the Social Fabric," in *The Barbarian Within* (New York: Macmillan, 1962), pp. 206–19; and *Fighting for Life: Contest, Sexuality and Consciousness* (Ithaca and London: Cornell University Press, 1981), pp. 129–34; and *Orality and Literacy* (Chapter 2, Note 25), pp. 112–15; and *The Presence of the Word* (Chapter 2, Note 28), pp. 249–52.

relationship between parents and children (especially adolescents) seems to involve a degree of it. It is almost as though the adolescent is a Fichtean ego who posits his or her parents as non-ego in order to be able to act and think. (Grandparents are usually exempted from functioning in the non-ego role with their grandchildren.) The non-ego is needed not just for growing up (which appears to be a dialectical business in our time) but also for narration.

The reliance on polarization in narratives means that the degree of conflict present in a situation or relationship gets exaggerated:[20] to tell something, and to render it more tellable and memorable in a retelling, is to heighten the element of conflict. The result is that the stories we tell about one another introduce some needless tensions into our lives. We can take measures to combat the exaggerations, of course, but it is important to realize that the amplification of conflict is part of the very fabric of a narrative.

Organized sports activities can help us come to terms with our combative impulses, for they show us how conflict can be viewed in less threatening terms. A great many sports activities require an opponent or an opposing team. Your opponent is often your good friend; yet for purposes of the sport or game, he becomes your enemy and does his utmost (within the rules, we hope, and not motivated by any personal malice) to defeat you. Part of growing up is learning to lose graciously at sports and games. To have learned this lesson is to realize that the person who played against you cannot reasonably be expected to have done anything other than what he did do, namely, strive to defeat you.

An opponent, then, is not necessarily an enemy[21] who wishes you harm or ill. Participation in sports can teach us not just sportsmanship (how to play the game fairly and properly) but also an attitude toward conflict which we all need, namely, that much of it is honorable and

---

20. This is one reason why it is a tricky business to tell stories to friends about disagreements in one's marriage relationship. If the stories get retold, the element of conflict is heightened, and soon people may be wondering: "Can this marriage be saved?"

21. This word shares an etymological root with "enmity."

inevitable, built into the human condition. The ego needs the non-ego in order to be able to function as ego. And after the game (or election, or contest), you and your opponent both have to go on living on this earth (and perhaps in the same community). You should therefore conduct yourself in a way that will make later coexistence possible.

Fiction and fact. The fourth constitutive principle of narrative consciousness and historical thought to be discussed is *integration,* by which I mean the tendency within true stories to seek a link with other true stories. All stories making a claim to historical truth are implicitly part of — or about — a single universe or framework of reference within which all the action takes place. To make such a claim is to distinguish true narratives[22] from fictional narratives. Whether such a distinction can be upheld is sometimes disputed.

The difference between fiction and fact has a great deal to do with the degree of closure that is achieved in a story. In the case of fiction, the universe posited in the narrative is closed, although under certain circumstances it can be opened again and expanded. In the case of factual or historical narratives, the universe of discourse always remains open. Therefore the story can, in principle, always be added to, supplemented, or joined with other stories to make a larger narrative.

Harry ("Rabbit") Angstrom is a character in John Updike's novel *Rabbit, Run* (1960). If I were to write a novel and include him, I would be in violation of the law, for he is a piece of literary property. The novel in which he appears contains many references to concrete realities, but its characters are fictional, that is to say, made up. Rabbit's world is closed, unless his creator should decide to open it again. He did so once when a movie was made on the basis of the book. And Updike also opened Rabbit's world when he wrote a sequel, entitled *Rabbit Redux* (1971). Some years later there came another Rabbit novel, entitled *Rabbit Is Rich* (1981), and then a fourth, entitled *Rabbit*

22. The use of the plural is deliberate. I would not wish to suggest that there can only be *one* true narrative about something that happened.

at Rest (1990). The four Rabbit novels are linked in the sense that they can well be read in sequence and then be understood as presenting us with a single fictional universe, a universe about which we can do research by reading and rereading the novels.[23] (Surely the later ones enrich our appreciation of the first one.) But it does not make sense to read the novels of Saul Bellow or Bernard Malamud, for example, in an effort to find out more specifics about the fictional universe inhabited by Rabbit Angstrom.

When it comes to historical narratives, we do not encounter the same degree of closure, for we read works by various different authors to see what light they can shed on each other. If a good book on the Vietnam war sparks our curiosity, we will probably read more accounts. We know that historians do not all agree and sometimes flatly contradict each other; yet as we read different accounts of the same events, we expect the stories to mesh, and in our mind there develops a larger sense of what took place. When we go through this process, we see the principle of integration at work: historical or real narratives link up with one another. The baton can be passed from narrator to narrator as in a relay race. Fictional narratives are sometimes seen as more entertaining because they are not subject to such disturbance. An author can write the story exactly the way he wants it, and then post a "No trespassing" sign, so to speak.[24] John Updike has a monopoly on the future foibles of Rabbit Angstrom, but no historian has a monopoly on the Vietnam war.

---

23. Robertson Davies has allowed Dunstan Ramsay, a character in his Deptford trilogy (made up of *Fifth Business* and *The Manticore* and *World of Wonders)* to pop up again in *What's Bred in the Bone,* which is the second volume of a later trilogy. One might form the impression that these two trilogies are part of a single fictional universe, but one cannot assume that all the works of any author must form a single universe. Such an assumption would not work for an author who writes both science fiction set in a distant future and novels set in the present.

24. Some history books have multiple authors, like the well-known Cambridge Ancient, Medieval and Modern Histories. Probably the most enjoyable history books are the ones with a single author, whose point of view and artful shaping of the material dominates and controls the whole narrative. When we prefer such works, it is essentially on aesthetic grounds; we are then assigning a lower priority to the concern for historical accuracy and fullness.

. . . we are never more (and sometimes less) than the co-authors of our own narratives.

–Alasdair MacIntyre[25]

No final version. What the individual who narrates his own history must recognize, in the light of the principle of integration, is that his own story is actually a part of *everyone's* story. Therefore he cannot have it to himself; rather, he may well see it swallowed up as it becomes part of a larger story. To the extent that he presents it as true, he must be willing to see it supplemented, and perhaps even corrected, by evidence and recollections that others can supply: an unwillingness to share the story and open it to scrutiny is a sign that one is not really committed to historical truth. For example, a man who saved someone's life years ago may not appreciate having the story retold by others who were also present if there is some question as to the extent of the danger he actually faced; he may be so eager to cast himself in the role of courageous hero (in his own mind, at least) that he avoids talking about the episode. We could say of such a man that he is guarding his memories so that he can "have it his own way." There are also people who mix a substantial dose of fiction into a personal diary written for their private enjoyment.[26]

The version of our story which we carry around with us and tell to others is also less than final in the sense that we can still add to it. Life is not yet over, and we do not know what future "chapters" will contain. Moreover, future events may place earlier events in a new (retrospective) light. Some statesmen and public figures try to close off their story by writing a substantial set of memoirs in which they foresee the important questions about their career and offer answers to them. They have every right to do so, and memoirs usually enjoy some success in shaping the way historians come to conceive of the story, but they, too, can be no more than provisional. No one has a monopoly on the truth about the events in which he has participated.

25. *After Virtue* (Chapter 5, Note 3), p. 213.

26. Patricia Highsmith's novel *Edith's Diary* (New York: Penguin Books, 1980) presents us with just such a character.

Anyone who wishes to declare his allegiance to historical truth must therefore allow for the possibility that his story will be substantially revised by the historians of the future. No wonder some people prefer anonymity!

Chapter 7

The Individual and the Group

Group truth? At this point in our discussion we need to confront the uneasiness felt by many people in the Western world when they consider the prospect of the "group" or society or some larger social unit playing a role in the process of establishing truth. Analyses of communism and other forms of totalitarianism have made it painfully apparent that groups and wholes can — and often do — thoroughly dominate individuals in the quest for truth. A communist interrogator in Arthur Koestler's novel *Darkness at Noon* says, "Truth is what is useful to humanity, falsehood is what is harmful."[1] And who is to decide what is useful to humanity? The Party, of course, and its leadership! The individual counts for nothing and can be so manipulated and pressured and confused that he will ultimately agree with *anything* the Party decides on as the truth.

To avoid going to extremes as we confront this question, we should ask ourselves whether it is possible for a group to act with integrity when it comes to the quest for truth. In principle I see no reason why it could not, although it might well be argued that a group is even more liable to corruption and sinfulness than an individual. Yet this is not to deny that an individual can be thoroughly deceitful and self-serving in his thinking and talking about the past. Neither the individual nor the group enjoys a priority or moral superiority over the other when it comes to the quest for truth. We must not allow the preference for the individual that is so widespread in English-speaking countries to carry us away, for if we disregard the input of the group

1. *Darkness at Noon* (New York: Macmillan, 1941), p. 227.

and of other people generally in the search for truth, we will have lost something of great potential value. Although no group can be said to possess a mind in the same sense that an individual does, a group can have memories, i.e. recollections which its members stimulate in one another. Those group memories are precious because the members of the group can cherish them in the company of the others: "Remember the time when . . . ?"

In considering the input of the group, however, it is helpful to distinguish truth *for* the group from truth that stems *from* the group. Totalitarians maintain that the group and its interests must always take priority over the individual: anything called truth must therefore be *for* the group and serve the interests of the group — at the expense of the interests of the individual, if need be. With this principle I would take issue: the rights and interests of the individual must not be trampled underfoot by the needs of the group. Now, I do not maintain that it is *easy* to achieve a proper balance between the interests of the individual and those of the group. It is possible to err in either direction. All I mean to do here is to repudiate the totalitarian position. The prospect I wish to hold out as a possibility is for the group, under ideal circumstances, to correct and supplement the individual in his quest for truth about the past — even about *his own past.*

When we observe families, we see that correction and supplementation do occur. A healthy family is a loving social unit in which the individual need not feel that his rights and interests are disregarded. As members of families we do look to others to help us find and build and maintain the truth about ourselves, our past, our life history. And we also seek to place our own story within the context of our family's story, thus using the family as a way to deal with the problem of where the story properly begins.

Consider the following narrative. Julia is looking back on a key episode in her own life, one which she was not able to understand properly at the time it took place, for she did not realize where the story began.

> ◘ One evening at supper — it was the spring of my grade 12 year — I announced casually that I would not be going to college after all. I had decided to get a job instead. Now, I hadn't actually *decided* this

> once and for all: I was really sending up a trial balloon. But my mother almost *exploded;* I'd never seen her so upset and angry before. Well, I wasn't going to get pushed around, and so I screamed right back at her. I told her that as far as I was concerned, the subject was closed.
>
> It was ten years later before I understood what had been going on that evening at supper. My grandmother (Mom's mother) told me that *they* had also had such a fight when Mom was in high school — but in reverse! Mom wanted desperately to go to college, but her parents wouldn't hear of it: they told her she would have to go to work and help support her younger brothers and sisters. (They had a large family.) Well, it seems that Mom decided that *her* daughter was going to college — no matter what! When I made my casual announcement, Mom lost her cool. If she had been more clever about it, I would probably have wound up in college after all. And if she had told me how her parents had held her back, I'm sure I would have been more sympathetic to her point of view. But she didn't — she came down hard on me instead. The upshot was that after our fight, I had to salvage my pride by *not* going.
>
> Well, I have two daughters of my own now, and I want *them* to go to college and use those opportunities I foolishly passed up. As I look back, I can see that Mom was right to get mad at me: I *have* lost out by not going to college. Yet, how do I know that the cycle of frustration won't repeat itself? Perhaps I should just tell my girls, when they get to be the right age to understand this sort of thing, about the fight between my mother and me, and the one between her mother and her, and what those fights were *really* about . . . .

This narrative illustrates the moral component in storytelling once again. In the episode with which it begins, there is much more in the picture than Julia realizes: her mother's anger and disappointment have roots in an earlier conflict; they are fueled by her painful awareness of her own lost opportunity, which she had never mentioned to her daughter. That lost opportunity is a better beginning for the story than "One evening at supper." Julia resolves to tell her own children a fuller story,[2] in the hope that historical truth will help her children

2. Not *the full story* in any strict sense, for it can never be told.

live their lives more fully. During her high school days Julia thought *she* was the one living her life — the only one. But an older Julia, a mother with two daughters, realizes that her story was also part of *the family's story,* and that her daughters must in some sense live part of *her* life, so that their story is at the same time hers.

Supplementation. When we allow other people — and perhaps some group — to play a role in the shaping of our sense of the past, we are recognizing the validity of supplementation in narrative consciousness and historical thought. The need for supplementation is grounded not first of all in any failings of memory but in the very *selectivity* of experience. On the visual level alone, ordinary experience presents us with such a wealth of texture and detail that we cannot possibly register all of it, to say nothing of focusing attention on every item and feature available to us. Thus, when we ask others to supplement what we remember, we are assuming that if they were with us as participants in the same event, they were aware of things that we missed. Also, they may remember things that we did experience but have forgotten.

The realization that supplementation does play a role in historical thought and in the construction of the stories we tell about our lives and experiences is helpful for getting us beyond a quasi-Cartesian conception of memory to which common sense continues to cling.[3] Our tendency is to suppose that our personal memory enjoys a kind of sovereignty, priority and independence such that when we think about the past, we can be *certain* only of that we which ourselves remember. The slogan for this conviction might be "Remembering is believing," which is parallel to "Seeing is believing."

3. I call it quasi-Cartesian not because Descartes actually propounded such a theory but because his philosophical method calls for us to doubt (if only provisionally) all that can possibly be doubted: "It is now some years since I detected how many were the false beliefs that I had from my earliest youth admitted as true, and how doubtful was everything I had since constructed on this basis; and from that time I was convinced that I must once for all seriously undertake to rid myself of all the opinions I had formerly accepted, and commence to build anew from the foundation . . ." *(Meditations* [Chapter 2, Note 15], p. 144).

The point of radical Cartesian doubt is to avoid having untruths imposed upon us by any sort of deceiver. Thus Descartes offers us his hypothesis of the "evil genius."[4] The proper use of the Cartesian method was supposed to guarantee that the individual could *personally* vouch for all that he claimed to know: nothing was to be accepted on hearsay or on the cognitive reputation of any other person. Now, the doubts Descartes raised were by no means foolish, for we are deceived quite often in everyday life. And so the recognition of supplementation as a feature of memory and narrative consciousness must include a sober warning that one cannot accept at face value *everything* that is offered to supplement our recollections and our awareness of what has taken place in the past.

In memory, as in much of human life, we depend on a network grounded in *trust*. The world of commerce could not function effectively if people did not trust one another,[5] and neither can the enterprise of remembering and maintaining a sense of the past. What we need to affirm, then, is that people can indeed be trusted — but not all of them, and not all of the time. The Cartesian approach goes wrong not in being too suspicious but in leaving us with a truncated sense of the past, of who we are and what we have been. In the long run it is worth taking the risk of trusting others to inform us about our past.

> *I felt that I could not write of*
> *these things unless I wrote the*
> whole *truth, not just a part.*
> –Czeslaw Milosz[6]

A single story. The difference between fictional and factual narratives, as we saw earlier (Chapter 6), is that the latter presuppose a single framework of space and time within which all the stories take

4. "I shall then suppose, not that God who is supremely good and the foundation of truth, but some evil genius not less powerful than deceitful, has employed his whole energies in deceiving me . . ." *(Meditations,* p. 148).

5. Edward Banfield has illustrated some interesting connections between trust and commercial success in his book *The Moral Basis of a Backward Society* (New York: Free Press, 1958).

6. *The Captive Mind* (Chapter 3, Note 2), p. 109.

place. As a result, we must think of the stories as able to be linked together and, indeed, as being strengthened through such links, for the stories both confirm and enlarge one another.

Yet the notion of a single story is essentially an *ideal*, one which we do not see realized to any great extent in practice. It may well be that Aztec civilization in Central America unfolded on the same planet as late medieval civilization in Europe, but we cannot speak of the two as being genuinely simultaneous. Therefore, in writing the history of the one, frequent references to the other would not make much sense, for they did influence and not interact with one another.[7] And on a human level, biographies of people widely separated in time and space do not, in practical terms, link up in the way that biographies of world leaders of the same era do.

Yet the ideal of a single story helps us see what historical truth can amount to, for in principle it holds out the prospect of progress. As stories link up with one another, individual bias and lack of perspective tend to become less of a factor. The larger and fuller story — at least, to the extent that it can be followed by finite minds — is presumably the truer story.

The notion of a single story has another meaning which needs consideration here, namely, agreement on the part of the widest possible set of people as to what sort of account of a given episode or train of events should be regarded as the true one. Yet, although the single story in this sense should be a truer story than the one being defended by some individual against all opposition, we should not press hard to see it implemented or made official. And we should not forget that sometimes a group of people under investigation for possible worngdoing meet to "get their story together": their intent is probably not to serve the cause of truth.

In a world without sin, a world in which people could always trust one another without reservation, it ought to be possible, in principle,

7. I do not mean to deny all exchange and interaction between civilizations in the Old and New Worlds prior to the time of Columbus; a similar point could be made about ancient Assyria and what we now call western Europe, even though there was no major geographical barrier between them.

to come up with "a single story." But because sin and corruption are present at all levels of our mental life, influencing experience, memory and narrative consciousness, it will not always prove feasible or advisable to come up with "a single story." Our democratic society allows people the right to dissent not just politically but cognitively; in other words, we enjoy the privilege of seeing and remembering things in a different way than our neighbors. Hence we will sometimes have to cling "our own story" and disavow the "single story" that people around us are trying to establish as the official and definitive account.

Overcoming centeredness. When people tell stories about events in which they were involved, they tend to make their own participation central and will generally heighten or magnify their own role. In such stories the narrator uses the first-person singular pronoun ("*I* did this, and *I* did that . . ."). Most narratives use the third-person pronoun to refer to the protagonist or chief agent ("*He*, or *she*, did this, and then that . . ."). But we should note that some people use third-person pronouns even when telling stories about themselves. In the final analysis, the use of pronouns is not a reliable indication of the degree of centeredness in a narrative.

There are a number of possibilities to be distinguished when it comes to my *degree* of involvement in the action which I narrate while making some use of the first-person singular pronoun: (i) I may be the *sole* agent in the story, (ii) I may be the *major* agent, (iii) I may be a *minor* agent, (iv) I may be a mere *spectator* reporting what I saw and heard, (v) I may be reporting what some *eyewitness* told me took place, or (vi) I may be reporting what I have *reconstructed* on the basis of physical evidence without the benefit of any testimony from eyewitnesses or participants. In some cases, then, I am personally involved in the action, and in others I am not involved at all.

That fact that we sometimes tell (third-person) stories about events in which we did not participate even though we did witness them indicates that it is possible to narratize in a mode that does *not* place the narrator at the very center of the action. Yet the tendency to make oneself central — or to exaggerate one's centrality or degree of involvement — is deeply ingrained. One of the ways we reach out for historical truth, then, is to try to overcome centeredness.

Organized team sports can help us understand the narrative possibilities here. When an athlete is asked to tell the story of how his team won the championship game, his own role in the game is likely to get prominent mention. If the game is basketball, it may well be that one person dominated throughout and was on the court for the entire game. In the case of football, we know that any one player, however important, will only be on the field about half the time, for in the higher ranks players do not play both offense and defense. Thus a football player, e.g. a quarterback, who falls into centeredness can easily be corrected. "Don't you think you should give the defense some credit too?" he may be asked. In the case of hockey, an account that maximizes centeredness is even more suspect, for hockey players (apart from the goalie) play in shifts. An outstanding defenseman may be on the ice for half the game or even a bit more, but there will surely be times when he is on the bench catching his breath. A forward will alternate in his position with two other players, and sometimes even with three. Thus no hockey player can make his account of the game the story of how *he* won the game: for much of the game he was on the bench. When athletes are interviewed, they generally manifest a healthy awareness of the dangers of centeredness, for they make much of the *team* effort. This emphasis on viewing things in terms of *the team* is one of the things that makes participation in sports educationally and morally attractive to many parents and teachers: children learn valuable lessons from team sports.

The temptation to lapse into centeredness also comes up in other group activities, such as warfare. When generals write about battles and wars they have been involved in, they often magnify their own role and importance. And the linguistic phenomenon of metonymy (see Chapter 6) also plays a role, as in such statements as the following: "Grant surged across the Rapidan and Rappahannock rivers on May 4, hoping to get through the tangled Wilderness before Lee could move. But the Confederate leader reacted instantly and, on May 5, attacked Grant from the west in the Battle of the Wilderness." In these two sentences, which are borrowed from an encyclopedia account of the U.S. Civil War (1861–65), we have a fair summary of one small slice of that war, but it is rich in centeredness. The account has Ulysses S. Grant, a famous Northern general, standing for the Northern (Union) army of which he was the commander, and then has him being attacked by Robert E. Lee, a famous Southern general, who stands for the Southern (Confederate) troops under his command. What really happened, of course, is that the two armies clashed. Like a sports event, the battle was surely a team effort on each side, but there is a tendency in military history to use the name of a general when referring to the deeds of his army: *Caesar* conquered Gaul, *Hannibal* crossed the Alps, and so forth.

The degree of centeredness one includes in storytelling depends on one's existential posture. In the posture characteristically assumed in youth, people tend to relate everything to *themselves* and to assume that things are important only to the extent that they themselves are involved as participants. For a small child, a toy is a thing of delight only if he himself plays with it: hence he rudely snatches it away from another child.

In old age, on the other hand, one gives up a certain degree of centeredness and comes to take much more pleasure in events in which other people are central. This is not to say that an older person becomes a mere spectator. Perhaps a comparison with an orchestra can help here: the older person recognizes that he need not play first violin in order to enjoy the performance and to feel part of the orchestra. This realization transfers into a willingness, on the narrative level, to tell stories in which centeredness is no longer exaggerated.

I do not mean to assert that all older people achieve the maturity and peace which such a transcending of centeredness manifests; some remain petty and self-seeking to the day of their death. But I do believe that some movement in this direction is the normal and proper pattern in aging. One comes to take genuine delight in the accomplishments and experiences of others, even younger rivals who are excelling in one's own major field of endeavor. A generosity of spirit sets in, and it manifests itself in one's storytelling. Not only is such a mature person able to acknowledge his own youthful foolishness; he is also able to give credit to others who played a major role in his own earlier successes, perhaps behind the scene. He may admit that his own parents were more of a force and factor in his life than he had earlier realized.

Ultimately, such awareness can lead to a revision of certain stories in the direction of truth. A man may say that he pursued a very special woman and made her his wife, only to discover later on that she was by no means a passive object in the process but had sought him even before he pursued her. Likewise, the older person may come more and more to see the truth of a hymn which Calvinists love:

> I sought the Lord, and *afterward* I knew
> he moved my soul to seek him, seeking me;
> it was not I that found, O Savior true;
> no, I was found of thee.[8]

> *A lot of facts never come to light. . . . Losers have their own truth.*
> –Julia O'Faolain[9]

Losers and enemies. In talking about losers, I am not just thinking of people who have lost battles; I also have in mind the people who tend to get lost, omitted, overlooked when the official histories are drawn up. If the group plays a major role in determining historical truth, losers are apt to be neglected. Here in North America I could point to the native peoples, who inhabited this continent for many centuries (and made much history) before white settlers arrived. Their history does not draw much attention, mainly because the native peoples are among history's losers: they were shoved aside in the great forward march we call progress. We even tell schoolchildren that "Columbus discovered America," which seems to suggest that the New World was uninhabited by humans before he got here.

History, by its very nature, is for the most part a record of success and alleged progress. Using a geographical analogy, we could say that it charts the highways that move on through time and lead somewhere — not the dead ends. Hegel, more than any other thinker, realized

---

8. Number 498 in the *Psalter Hymnal* (Grand Rapids: CRC Publications, 1987). Italics mine –TP.

9. *No Country for Young Men* (Chapter 2, Note 1), pp. 50–1.

this and quoted Schiller's famous line about world history being the world tribunal.[10] Since historical understanding is retrospective, it cannot help but read like a report drawn up by the victors, a report that does not take time to shed tears for the losers.[11]

The single story ideal suggests that eventually the winners and the losers should be able to get together and agree on an account of what happened. Would such a thing be possible in connection with a monumental conflict like World War II? In principle, yes, although we should not expect much while the combatants are still alive. Surely it can be said that a measure of agreement has already been reached between the various participants as to *what* happened and *why*. Yet the losers in warfare are not always heard and respected. How can we do more to ensure them a voice and a remembrance?

It is not enough to encourage the losers (i.e. victims, those who suffered and were betrayed) to tell and record and publish their story: the physical evidence (traces of the past relating to their fate) also needs to be preserved. Naturally, not everything can be kept just the way it was; we must be willing to rebuild and go on with life. But surely the decision not to re-use the land on which some of the concentration camps and death camps were built was a wise one.[12] Also valuable in this regard are museums and other displays.

We also show respect for history's losers when we recognize that they are not necessarily moral losers. Here Hegel haunts us again, for he tended to fuse historical success with moral success. In many cases, the oppressed, who may have been beaten down to historical insignificance, are essentially victims. And although their story leads nowhere,

---

10. "Die Weltgeschichte ist das Weltgericht." See *Hegel's Philosophy of Right*, tr. T. M. Knox (London: Oxford University Press, 1967), Section 340, p. 216.

11. Hegel spoke of history as "the slaughter-bench at which the happiness of peoples, the wisdom of states, and the virtue of individuals have been sacrificed": see *Reason in History* (the introduction to his lectures on the philosophy of world history), tr. Robert S. Hartman (Indianapolis and New York: Bobbs-Merrill, 1953), p. 27.

12. Such sites are only an approximation of what once was, even if they do represent the exact geographical location. The full horror, including the stench, of what used to go on, could be maintained only if the camps kept operating!

historically speaking, it needs to be included in the larger historical account, even if its presence in the record is a vexation for the triumphant majority. To put this commitment to historical truth into practice requires courage.

What if the historical loser is also a moral loser, as in the case of Hitler and the Nazis? Here, too, there is a minority report for which we will have to make room, even if we disagree with it. Some ordinary people in Germany today maintain that although Hitler did wrong, he was nowhere near the villain that the victorious Allies made him out to be. And there are historians who debate this question in a responsible way, pointing to certain accomplishments for which Hitler did not get enough credit. The writings of Albert Speer[13] have also led people to reconsider certain aspects of the Nazi story.

But what if the "minority report" includes the claim that there was no Holocaust, no murder of millions of Jews and further millions of non-Jews? This hotly debated question has even been before the courts in Canada[14] and also affects the schools.[15] My own view is that the ideal of a single story should not override the rights of minorities to hold — and publish — their own opinions (however unsubstantiated) on historical issues.

But when it comes to teaching those opinions in schools, the local school board must make some decisions. No teacher is free to teach absolutely anything he likes; he is not free, for example, to teach something that all his colleagues in his own field of expertise regard as scientific or academic or historical nonsense. Schooling is a *corporate* enterprise. Thus, if there were such a thing as a Nazi school

---

13. Speer was a high-ranking Nazi leader who survived a long imprisonment after the war and then wrote about the Nazi era and his relationship with Hitler in his book *Inside the Third Reich: Memoirs,* tr. Richard and Clara Winston (New York: Macmillan, 1970).

14. In 1985, Ernst Zundel, a publisher in Toronto, was tried and convicted for putting out material that denied the Holocaust.

15. In 1982, Jim Keegstra, a teacher in a public school in Alberta, was dismissed for denying the Holocaust in the classes he taught. He also lost his teaching certification and was eventually brought to trial and convicted on a charge of "fomenting hate against an identifiable group." See David Bercuson and Douglas Wertheimer, *A Trust Betrayed: The Keegstra Affair* (Toronto: Doubleday Canada, 1985).

in Alberta, it would surely teach a view of World War II that the vast majority of Albertans would find utterly abhorrent, but people outside the school would not be in a position to do anything about it. The only real alternative to such a position is to grant the government the authority to determine what is true and false in history and the sciences and to enforce its view of history in the content of all schooling. A government with such powers would come dangerously close to outright totalitarianism.

Relevance and significance. When we turn to the place of the little people in history, including individuals who have made no name for themselves and are likely to be forgotten, a distinction between relevance and significance can help us. The group or larger social unit may be right in ruling that various individual lives and episodes have no place in the collective history simply because they are not "important" enough — provided that in this case we understand importance in terms of *relevance.* There are indeed people whose lives have run in such ordinary and peaceful channels that they attracted little notice outside their immediate family and employment circles, people who became neither famous nor infamous. Such people are not included in the continuing official history which the group or community maintains regarding itself and teaches to new generations of children. Surely there is no historical injustice in this.

Yet it is also possible for people and their lives to be important in a more fragmented, individual sense: they are important to *themselves* and to their families and loved ones.[16] It is widely accepted in our Western world that every life is significant and valuable in and of itself: we repudiate the communist emphasis on the group, the larger whole, for this emphasis leaves the individual out of the picture.[17]

---

16. Ranke's claim that every epoch is "immediately present to God" (see his *The Theory and Practice of History,* ed. Georg G. Iggers and Konrad von Moltke, and tr. Wilma A. Iggers and Konrad von Moltke [New York: Irvington Publishers, 1983], p. 53) could perhaps be broadened here to include every *individual human being* as important in God's sight.

17. "The individual was nothing, the Party was all; the branch which broke from the tree must wither . . . ," writes Koestler in *Darkness at Noon* (p. 81). The indi-

To deal with this issue of significance in relation to historical truth, it will be helpful to distinguish between "famous" and "historical." That a certain person is not famous, i.e. widely known in his own society, is a simple matter of fact, even if some people were firmly convinced that he *deserves* to be famous. But someone who is not famous does not thereby lack a history: what we would say of him instead is that he has not made it into the standard and widely read history books. However, there are also *non-standard* histories; all members of a free society are allowed to write and distribute such histories.

Linked with this commitment to pluralism in writing and publishing there should be a commitment to pluralism in education. Minority groups in Western society are free to write and publish and teach their own histories, even if those histories are deemed irrelevant to the larger history (in the sense explained above).[18] Pluralism of this sort is not an ideal state of affairs: its tenuous existence in our society[19] runs contrary to the ideal of a single story. But much greater social harmony and cultural integration than we currently see in our society would have to be achieved before "a single story" could be taught in all schools as "history." In the present divided situation, the "winners" should try to be gracious to the "losers" and leave them room to develop and teach their story.

Prehistory. Another reason why "a single story" cannot now be written and taught to all children in all schools in our society is that differences regarding prehistory must be taken into account. By prehistory I mean not just what happened in the period before mankind left written records behind (the usual definition) but the entirety of the early sector of man's sojourn on earth which cannot be explored

vidual was not understood in qualitative terms: "The definition of the individual was: a multitude of one million divided by one million" (p. 257).

18. My own interest in this issue is fueled partly by the fact that I spring from a rather small ethnic minority, i.e. the Frisians, who live along the North Sea.

19. There are *public* schools in which the "majority report" view of history is passed on to children; fortunately, parents are free *not* to send their children to such schools. I am a proponent and supporter of independent Christian schools.

by normal historical methods. In other words, I am also referring to the very beginning and to those periods for which there are no artifacts, no cultural "traces of the past." In Biblical terms, this is roughly the period about which the first eleven chapters of the book of Genesis speak.

The Christian doctrine of creation is a piece of prehistory, for it is not based on ordinary historical sources or records from which we surmise that such-and-such must have happened, nor does it result from our analysis of a process that has left traces behind. If the world was indeed created *ex nihilo* (out of nothing), there will always be something unfathomably mysterious and abrupt about how things came to be; we need not expect that some clever scientist or historian will figure it out one of these days. However much we may speculate about how the world might once have been as we trace processes and developments and as we engage in retrodiction,[20] we run up against a blank wall when we come to the very beginning and find that we can inquire no further.

Because the doctrine of creation is a deliverance of revelation, it will not be accepted by people outside the Christian tradition (at least, not in its Biblical form, although there are also creation accounts in some other religious traditions). This doctrine is the perpetual prologue to every Christian story — both individual and collective. This prologue assigns to man his place in relation to God, the animals, and all creation. He is given a role and function on the earth. And in a narrative that follows on the heels of the creation account we get another piece of important prehistory, one that locates the origin of human suffering and of impulses toward evil in an act of rebellion against God. Again, this narrative does not have ordinary sources; our knowledge of it is derived from revelation. Yet the traces of what happened seem written into human nature, with its tendency toward deceit and corruption.[21]

---

20. The opposite of *pre*-diction. To pre-dict is to say in advance what will happen, and to retro-dict is to declare after the fact what must have happened.

21. From man's longing for an ideal existence and his capacity to imagine a paradisal state, one can almost infer that he once lived in such a state.

Our narratives need prehistory by way of background and prologue because our stories have a *moral* function in our lives. Not only do we use stories to justify ourselves and our conduct on certain occasions; we also use them to orient ourselves and to stir up within ourselves the moral courage (think of the word "encouragement") which we need in order to shoulder our extensive responsibilities as human beings. The Biblical prologue to our story tells us that we are created in God's own image.

It is significant that the Biblical revelation concerning God's image in man is given to us in the context of the creation story and not, for example, in the New Testament. Its placement so early in the Bible suggests its significance for man as a maker of history. Biblical prehistory *encourages* us, then, and this is one of its substantial differences with evolutionism, one of its major rivals.

The Darwinian picture of human origins also includes a substantial dose of prehistory, but it is of such a nature as *not* to provide man with much hope or encouragement; indeed, part of the failure of the Darwinian vision in the twentieth century has been its inability to articulate broad-ranging moral imperatives that would help set a direction for social and cultural life. So-called social Darwinism in the nineteenth century did not have such an effect either and was somewhat debilitating for the general run of people who came to believe in "the survival of the fittest." The demoralization we find in the western world today has a good deal to do with Darwin's influence.[22]

Eschatology. While we are discussing the question of (re)moralization, we should also consider eschatology (the things to come, or "last things") as a counterpart to prehistory. While we cannot be certain what the future will bring, e.g. whether it will frustrate and cast down what we have labored to build here on earth, the Bible does contain some teachings on this matter; unfortunately, they are not

---

22. What we need, it appears, is "remoralization," which is a term I borrow from Arthur Kleinman's exploration of narratives in the lives of people who are chronically ill. See *The Illness Narratives: Suffering, Healing and the Human Condition* (New York: Basic Books, 1988), especially pp. 54, 97, 237, 244ff.

sufficiently clear as to produce agreement among all Bible-believing Christians. There are various traditions when it comes to eschatological thought, and the different points of view have implications for how we conceive of the bigger story — ideally, the "single story" — to which all human endeavor contributes.

The forms of eschatology usually called *pre*-millennial[23] stress discontinuity between human history (including the labors of believers) and what is to come. The return of Christ is in effect an interruption and comes when things are in a deplorable state on earth. Christ inaugurates his thousand-year reign, although in some schemes there are a number of events before that reign actually begins. The effect of such thinking is to discourage efforts to "build the kingdom of God on earth," since we know in advance that it will come to nought.

The eschatologies called *post*-millennial, on the other hand, take a more hopeful view of human cultural endeavor: our labors serve to prepare the way for Christ's perfect rule and will inaugurate a golden era of a thousand years which will *precede* Christ's return.[24] This outlook has the effect of encouraging Christian cultural activity and endowing it with great significance. It tends to link God's story (what God is doing) with our story much more than any pre-millennial view does, thus leaving us with a more compact and tidy overall picture of history. As a result, it encourages the development of a somewhat more dogmatic "Christian view of history" — more dogmatic in the sense that it claims greater certainty about what God is doing in the world than pre-millennial thought can claim.

As we consider prehistory and eschatology, we see again why the notion of "a single story" remains an *ideal;* it is not a working model that we can apply to everyday historical thought as we deal with specific events. Christians today are not as certain as Christians of earlier eras about what the first eleven chapters of Genesis mean in terms of time: it appears that the divine record does not provide us

---

23. I use the plural (*forms* of eschatology) because there are many versions of pre-millennial thought.

24. The "pre" and "post" distinction has to do with the time of Christ's return: will he come *before* the thousand-year period (*pre*-millennial) or *after?*

with much chronological detail regarding events between Adam and Abraham. Thus, if there is, in principle, a single story, we are not now in a position to be able to tell it in any detail; we possess, at best, a sketchy version of it. And even if one should hold quite confidently and dogmatically to a detailed eschatology of either the pre-millennial or post-millennial variety, it still has to be acknowledged that a great many Christians do not accept all the details of that eschatology. Thus, with regard to what lies ahead, there is some uncertainty.

A Christian view of history needs prehistory and eschatology as parts of the total picture, but it uses them mainly to encourage and remoralize us for the task at hand. It does not regard them as a detailed map showing us how we got to where we are or as a specific itinerary announcing where we are headed. It appears that God has some surprises in store for us.

God's rule in history. Reformed Christians generally do not hold to either a pre- or post-millennial eschatology. The view they espouse instead is often called a-millennial.[25] In more positive terms, the standard Reformed position (which admits of some variations which will not concern us here) could be called the "kingdom of God view," for it stresses the reign or rule of God during the thousand-year period (symbolically understood as a period of fullness) between Christ's first and second comings. God maintains his sovereignty, on the one hand, but he does allow the devil and the forces of evil some room to operate, on the other hand.

This eschatological vision is the background to a Reformed understanding of the Christian life in this world. On the one hand it encourages and remoralizes us, but on the other hand it does not allow a facile optimism and triumphalism, for Satan is still to be reckoned with, as our everyday experience also suggests. It lends meaning to our daily activities but cautions us not to be presumptuous regarding the

25. The issue between these three is what is meant by the thousand-year reign of Christ referred to in Revelation 20. The a-millennial view does not interpret this period as a literal thousand years — hence the "a" prefix, which negates the term that follows.

significance of the labors of our hands, for they can so easily be undermined or destroyed by the forces of evil. Instead the Reformed eschatology leads us to utter an ancient prayer: "Ora et labora" (Pray and work).

Our eschatology encourages us especially on dark days when we wonder whether we are losing the battle here on earth. If, like Elijah, we want to throw up our hands and exclaim that there is no one left who is still faithful (see I Kings 19), our eschatology reminds us that when history has run its course it will be seen that we are *not* the losers after all but the winners. Now we know the story — past and future — only in part, but what we do know of the final outcome gives us the strength and conviction to carry on even when the sky is black. In the final analysis, there is a single story to be told, and in that story our own life, however insignificant it might seem at times, finds its fulfillment.

Chapter 8

Refinement and Revision

When half of a marriage packs up and goes, it takes the past with it. The present no longer reinforces memory, it casts doubt on it.

–Violet Weingarten[1]

Presumption of truth. The system of administering criminal justice in most Western countries operates by the principle of the presumption of innocence: someone who is suspected of committing a crime is *presumed* to be innocent unless and until he is found guilty. This principle is widely thought to be basic to the freedoms we enjoy and is staunchly defended. Yet this presumption in favor of goodness (understood in a general sense) does not seem to extend into the analyses of knowledge which philosophers offer, for many of them tend instead to regard candidates for belief as guilty (i.e. erroneous or false) until such time as they are *proven* innocent, i.e. true, or perhaps credible, worthy of acceptance.

Skepticism is a much-discussed theme in philosophy, for many of the thinkers who have explored human knowledge have been deeply impressed by the possibility of illusion. Some have concluded that our human sensory apparatus cannot be trusted, and that we must virtually regard error as normal: our perceptions are not to be accepted as truth unless they are somehow validated by a process which is not part of immediate experience. But others have claimed that a valid and adequate apprehension of reality is to be regarded as *normal,* and that illusion is the exceptional case that needs explaining. A Christian philosopher would approach this issue by arguing that since our

1. *Half a Marriage* (New York: Pocket Books, 1977), pp. 52–3.

sensory apparatus is given to us by God and is intended to guide us in our life in this world, we should be able to rely on it under normal circumstances.

This debate also carries over into philosophy of history, where memory takes the place of perception as a primary source of knowledge. The question is then whether we can trust the primary narratives[2] arising from memory, as opposed to regarding them as illusory and likely to mislead us. Without embarking on a discussion of perception, I will maintain that memory should be regarded as enjoying the presumption of truth. In other words, when we have no reason, under normal circumstances, to suppose that we are confused, we should regard what we remember as actually having happened. If I am reading in my study one afternoon and remember that an old friend dropped by unexpectedly for coffee that morning, I have no reason not to regard that memory as true and valid.

This is not to say, however, that memories and the narratives to which they give rise cannot be improved upon. When it comes to memory and history, truth is not a black-and-white matter.[3] I might think back to the conversation with my friend who dropped by for coffee, perhaps while discussing it with my wife, and see new angles in it, making new connections between remarks he made. In that sense I can improve and refine my memory of it. And I may also supplement it if my wife adds something I had missed, perhaps because I was not paying attention to the same things she focused on.

Memory in this ordinary sense is basic to learning. Students who have attended a philosophy class remember something of what the professor said. Presumably they made some notes to consult later. If they review those notes the same day and also call up what they remember about the class and reflect on all of this, their memory of the lecture will be substantially improved. They can refine their mem-

---

2. See the Glossary.

3. One characteristic memory error is that we "misplace" what is otherwise a valid memory: I remember a dispute clearly and vividly but wrongly identify my antagonist, or I remember a chess match well but place it in the wrong year. In such cases we have a *mixture* of truth and error.

ory ever further by discussing the lecture with classmates. Sad to say, many do not take the time to do any of these things. One reason for their negligence in this regard is that they tend of think of memory as inert. Just as a tape recording does not improve itself some hours after it was made (it simply records and retains certain sounds), so, they imagine, their recollections of the philosophy class sit unchanged somewhere in the brain, waiting to be recalled for future use at test time. But memory is alive: it refines, organizes and compacts its contents.

These everyday examples show us that the (normally) true memories and narratives we carry around within us are subject to a process of refinement and revision. The old-fashioned word "amelioration" comes to mind here: something that is already good (e.g. my memory of my friend's visit or a student's memory of the philosophy class) can be improved, made better. This is in good measure what goes on when people "do history"; that is to say, they construct and then improve narratives which are supposed to embody truth about the past, drawing on traces of the past, memory supports, and the input of other people.

> *What I have written here strains to be true but nevertheless is not true enough.*
> –John Updike[4]

Contested truth. The way truth is pursued in the natural sciences may also shed some light on the nature of narrative consciousness and historical truth. The first point to be made here is that observations, whether in ordinary experience or in the controlled world of the science laboratory, do not *dictate* hypotheses or theories; rather, hypotheses are generated via a process that presupposes a (provisional) theoretical framework and also draws on imagination. Hypotheses are then tested against observations and experiments that have the potential to disprove them. Those that fit best and turn out to be the most productive (in both theoretical and practical respects) are provisionally accepted as true, as the best that is currently available.

---

4. *Self-Consciousness* (Chapter 2, Note 21), p. 234.

This understanding of science, which is associated especially with the thinking of Karl Popper, emphasizes revision more than refinement. Popper does not maintain that science gets ever more accurate, refining and polishing its hypotheses until they become as hard as diamond — so hard that they will withstand any conceivable pressure from new observations or theories. Rather, hypotheses and theories succeed one another: the one that works best is king, so to speak, until another one comes along to take its place because it is even more effective. The new theory is often a revised version of the one it replaced. The upshot is that scientific theories do not get *confirmed* (proven true once and for all); instead they are *falsified.* The "true" theory is one that has not been falsified.

What about the dethroned and discarded hypothesis? Are we to look back upon it as having been worthless, an illusion? The usual answer is that those who held it were by no means foolish to do so — it was the best they had. And the discarded hypothesis may well have a fair amount in common with the one that took its place. Thus there is an element of continuity in the process by which scientific theories succeed one another.

The search for truth in history also contains an element of revision in Popper's sense: there is more to it than refinement. Moreover, just as there is a *contest atmosphere* to some scientific research (teams of scientists try to overthrow an existing theory by showing that a new one they have developed makes better sense of the observations and experiments, including some new ones they have undertaken to buttress their position), so historians try to unseat one another's interpretations and engage in considerable "revisionism." It is often a good exercise for a young historian to try to amass evidence against a widely held historical theory or interpretation, just to see whether any sort of case can be made against it.

Thus the process by which historical truth grows resembles a contest, complete with prizes (academic honors of various sorts). Just as we encourage free enterprise in the marketplace, in the belief that it will be best for all concerned in the long run, we encourage a rivalry and competition in scientific and historical research. Complacency is discouraged: any historical explanation or interpretation can, in principle, be challenged.

The comparisons I have made with the philosophy of the natural sciences bring to mind the thinking of Thomas Kuhn,[5] whose influential view of how changes come about in scientific theorizing differs from that of Popper. Without committing myself on the issues in philosophy of science that distinguish Popper's position from Kuhn's, I will argue that because historical thought and interpretation is so much closer to common sense than the kind of thinking that goes on today in the natural sciences, the Kuhnian model of a "paradigm shift" or a "revolution" in the entire framework of an academic tradition or discipline does not apply to history. More particularly, although a Kuhnian shift in historical thinking and interpretation would be *possible* — at least in principle — it would have to take the world of ordinary experience with it, and we see no signs of any such thing happening. Hence I maintain that Popper's model comes closer to giving us a sense of what transpires in the clash of historical interpretations than Kuhn's approach. There is indeed a contest between rival historians, but they *do* understand one another's arguments, unlike Kuhnian antagonists who ultimately bump up against the barrier of "incommensurability."[6]

> *But that's the wonderful thing about* fiction . . . .
> *You can rewrite it and rewrite it until it satisfies you.* You *have control. . . . Unlike real life, which can be so obstinately uncooperative.*
> –Nancy Thayer[7]

Primary narratives. The set of narratives that lives in conscious-

---

5. For Kuhn's views, see *The Structure of Scientific Revolutions,* second edition (Chicago: University of Chicago Press, 1970).

6. Kuhn maintains that when there is a revolution or paradigm shift, the thinkers who adopt the new framework of reference cannot be understood by those still working within the old one: the two positions, along with the claims made within them, are incommensurable. In many cases, the spirited debate between holders of competing positions which outsiders to science might expect to see does not take place.

7. *Morning* (New York: Charles Scribner's Sons, 1987), pp. 120–1.

ness is the raw material forming the point of departure for the historian, or for anyone who takes a reflective interest in the past, whatever his academic qualifications or his day-to-day work. Those narrative recollections, which are to be distinguished from other sorts of recollections we retain as well, get linked and refined. Sometimes the resulting narratives are revised in a substantial way, which usually means that there were lies or falsifications — or at least significant errors — within some of the primary narratives. Such, in brief, is the process by which we seek and build historical truth.

Some people might argue that "the facts" form the raw material or starting point for historical thought. Now, if this were the case, the past would not be quite so elusive: it would stand before us concretely, for the so-called facts people appeal to are all items in *present* experience. Some of them are statements about the past that are never called into question. To call them facts (in the sense that one could appeal to them in an argument without fear of being challenged) may well be justified, but they are still *present* realities, that is, contents of our consciousness today. Facts in this sense do play a role in historical thought and argumentation, but we cannot build historical research on them alone: the point of such research, surely, is to explore what is *not yet well known* — in other words, that about which no consensus has formed. If we want to write the history of an obscure and seldom discussed battle in World War II, for example, it will not help a great deal to find out what all the standard histories and encyclopedias say, for they do not contain much about the matter. Rather, the place to start would be with the consciousness of the people who lived through that battle and have some recollection of it which they can turn into a narrative or set of narratives which the historian could in turn proceed to test and refine.

When people speak of "the facts" as the place to start, they might also mean the traces of the past, which we usually call historical *evidence.* Such evidence does play a major role in historical inquiry, just as it does in the investigation of criminal activity, which is also historical insofar as it deals with what happened in the past. Yet traces of the past do not by themselves generate a narrative. Where there are no recollections or written sources embodying recollections of people who are now dead, it is indeed possible for a historian to study evidence and generate a narrative of what must have happened. The narrative produced in such an instance is not dictated strictly by the evidence on hand: it also draws on the historian's imaginative capacities and on various assumptions about human behavior and about the way things usually go in life. But in cases in which there are still some narrative recollections, the usual procedure is to start with them. The historian either refines and supplements them or, if they appear to be unreliable, discards them.

> *. . . in order to remember certain things that happened six years ago, I had to make a real effort, not so much to remember them as to imagine them. Just as when you're talking about another person and, on the basis of some piece of information, you make hypotheses about how certain events came to happen.*
>
> –Alberto Moravia[8]

Bypassing memories. The widespread tendency to suppose that one could build narratives about the past *directly* on evidence, that is, on traces of the past, is probably the reason why philosophers of history have, by and large, ignored memory. If it were possible to bypass memory and work directly from evidence and still come up with an adequate picture of what happened, we could safely ignore people's recollections and the memoirs of the statesman. Yet there are cases in daily life when we have no recollection of some episode but still wind up generating a narrative of what must have happened, using records and traces of the past as our basis. We may then be left with a recollection which later seems to us to have been a narrative recollection all along. In other words, we are liable to forget the process whereby we (re)constructed what seems to have happened. Memories do not bear intrinsic marks of their own authenticity.[9]

---

8. *The Lie,* tr. Angus Davidson (New York: Farrar, Straus & Giroux, 1966), p. 54.

9. Elizabeth Loftus presents an interesting example of a prompted, constructed "memory" that seemed authentic but was not: see *Eyewitness Testimony* (Chapter 3,

To understand the role of facts and evidence, we must recognize that a narrative — or at least some provisional sense of what must have happened — is needed in order to give the evidence being used a meaning and a place in the inquiry. A farmer's field which was the site of a small, little-known battle in World War II surely represents historical evidence, but an inspection of that field by a historian will reveal little that is of historical interest if he is not already inquiring into that battle. The historian needs to bring *questions* to his inspection of the evidence. If he has a narrative hypothesis in mind, he might study the landscape, perhaps making some assumptions about the size of the trees at the time the battle took place, and try to determine whether his provisional sense of what happened will stand up in the light of the topography he finds. But he will also be careful to inquire whether the landscape has been altered in any substantial way in the years since the war. Could there once have been a creek that was later rerouted? Was there a little-used railway line along the field that was later torn up? The field as it now exists is not a pure piece of the past: it is indeed relevant to our inquiry into the past, but it needs to be "interpreted" and investigated with an eye to the historical problem.

Although recollections and narratives enjoy primacy in historical inquiry (since without them there would be virtually nothing to talk about), the actual evidence is of great importance in the process of refinement and revision. The evidence *constrains* the narratives we recollect and construct. Fiction can be rewritten wholly at its author's whim, so that the story comes out exactly the way the author wants it to, but the writer of a factual and historical narrative must always bear the evidence in mind. If his narrative is to have any hope of being accepted as truth, it must comport well with the evidence in existence — unless the historian is prepared to attack the legitimacy or authenticity of the evidence. (History acquaints us with many examples of faked and discredited evidence.) Our freedom in constructing narratives about the past is limited by the recognized remnants of the past,

---

Note 7), pp. 62–3. The person whose memory turned out to be radically mistaken in her example was the Swiss psychologist Jean Piaget.

and in that limitation we have a valuable resource when it comes to seeking truth in historical awareness.

> *... history is not the writing but the* rewriting *of stories ....*
>
> –Louis O. Mink[10]

Academic history. We are now in a position to see what history, understood here as a professional undertaking carried on by trained specialists, aims to achieve. The people who devote their life to the work of historical research and the writing of history are not engaging in a kind of thought that is qualitatively different from that in which a normal human being engages when he looks back over his own life and tries to find order and continuity in it. Of course a trained historian has a more sophisticated grasp of the elements we draw on when we reconstruct the past, but his result is normally a narrative (or the revision of a previous narrative) that is of such a nature that ordinary people can follow it. Thus we find that people who have no higher education or specialized training to speak of nevertheless read the writings of academic historians. Yet it would not occur to them to try to decipher what physicists and biologists write in their professional publications. The writings of social scientists fall in between. Some do write in a way that communicates effectively to non-specialists, but their work tends to be dismissed by colleagues as "popular" — not sufficiently rigorous or scientific. Often journalists, who know how to stay away from technical terminology, manage to communicate the insights developed by social scientists.

One consequence (if this view of written history is correct) is that we should not think in terms of some people as being "historical," i.e. belonging to the set of people and events in which historians are interested, whereas others are not. It is true that not everyone makes his way eventually into a history book, but whether one does or not is in good measure due to the choices made by historians. And one's inclusion in mainstream academic history should not simply be regarded

10. "Philosophical Analysis and Historical Understanding," in *Historical Understanding* (Chapter 5, Note 4), p. 136.

as a badge of merit, for people do make it into "history" for evil deeds as well as for laudable ones: Lee Harvey Oswald was a "nobody" by many standards, but when he opened fire on President Kennedy he carved out a lasting remembrance for himself in our history books.

Communities and minority groups create and establish historical importance when they investigate their own history and commit their story to paper. As they do so, they confer historical recognition upon their leaders, even if mainstream academic history ignores those same leaders. It is open to anyone to write and publish a biography of his own father, thereby endowing him with some historical permanence (just as it is open to all of us to decline to read that biography). Professional historians are also engaged in this business of drawing our attention to largely forgotten figures in the past: one of their characteristic projects is to demonstrate the importance of some person or group or movement that has been ignored.

Which people make it into history as taught (which is not the same as history as written) depends on the decisions of those who set the curriculum in our schools and also on those who do the actual teaching. Many figures in the past have been written about to some extent but are no longer covered in courses and therefore gradually sink into oblivion. It is open to a religious or ethnic community to establish schools of its own with history courses in which figures within the community's past and tradition get covered and remembered: this is one of the major reasons for the establishment of independent schools that are religiously based.

But what takes place in higher-level history teaching is not just a survey of narratives considered important: much of it is also criticism, reconsideration. Higher-level, academic history normally takes primary and received narratives as its starting point and then proceeds to criticize them, rewrite them, and in some cases cast them aside as false. Often the result is a new narrative, which is deemed by its developers to be more accurate. Louis O. Mink writes: "In many cases . . . historical research is undertaken for the sake of criticizing and revising a received or orthodox narrative or an interpretation whose very formulation is displayed in particular narrative structure. In the process of revealing narrative gaps in such a narrative, or of showing its incapacity to assimilate relevant incidents or data, an alternative story may gradually take shape and undergo elaboration and revision, in turn suggesting new questions for research."[11] Arguments over possible alternative narratives covering key episodes in history lend interest to a many a history class taught on the post-secondary level. (On lower levels, the tendency is to stick to the basic narratives — or "the facts," by which is meant the widely accepted view.)

Academic history, in both the writing it produces and the teaching it does,[12] also serves to make people aware of the structure of the thought processes by which we keep the past alive in memory and historical consciousness. (Many of the issues dealt with in this essay form part of the content of higher education; they are matters on which educated people need to reflect if they are to function effectively as citizens of a democratic society.) Thus there is more to teaching history than telling — or reminding — people about what happened in the past. Historians also take the lead in showing society how to preserve appropriate records and traces of the past.[13] Those records and traces are important for guaranteeing the possibility of historical truth, for without them future historians will not be in a position to mount a valid challenge to widely accepted narratives which may be inaccurate.

Overriding descriptions. Some thinkers who discuss historical knowledge see a greater gulf than I do between professional historical thought, on the one hand, and common sense thinking about the past, on the other. Their position is that the historian frequently uses overriding descriptions; that is to say, he sets aside the agent's own ac-

---

11. "The Divergence of History and Sociology in Recent Philosophy of History," in *Historical Understanding,* p. 178.

12. A great many academic historians are teachers or professors in colleges and universities.

13. This is not quite so simple a matter as one might suppose at first, since there is much more material of potential historical interest and value than our archives and museums can possibly contain. And some of the great battlefields of history may have to be turned into farms or shopping centers, since land is a valuable resource.

count of what he was doing and replaces it with another of his own devising, one which the agent would not accept as correct — and perhaps would not even understand. Marvin Harris defends a version of this approach when he contrasts "emic" analyses, which agents supply as accounts of their own actions, with "etic" analyses, which the academic observer develops and regards as superior.[14] According to his theory, the agents do not know exactly what they are doing or why they are doing it — but they manage some success all the same![15]

People often object to such overriding descriptions because they do not seem to manifest much respect for the human person, for subjectivity and intentionality. (Harris indeed describes his position as "cultural *materialism*.") But we need not reject the notion of overriding descriptions simply because we wish to maintain a "high" view of man. Surely it is possible to uphold that some people really are like the ones Harris discusses in his sketches: their life includes patterns and customs that make sense, but they themselves are not aware of why. In other words, they are not able to explain why the things they do work so well for them. Full self-consciousness, by which I mean a full awareness of the rationale behind the things we do, including the customs and traditions we have picked up from our forebears, is not to be found among human beings. A good measure of self-consciousness and deliberateness is something to strive for, but there is nothing degrading to man in the notion that our traditional ways of doing things embody a wisdom which we are not able to articulate fully.

Another version of the use of overriding descriptions can be found

---

14. Harris is an anthropologist and offers us methodology to be used in social scientific research. But the contrast he draws can also be applied to history. For his account of the distinction between "emic" and "etic" explanations, see *Cultural Materialism: The Struggle for a Science of Culture* (New York: Random House, 1979), pp. 32–41. Harris shows the appeal of his approach in some of his other writings, in which he offers overriding descriptions of curious customs: see especially *Cows, Pigs, Wars, and Witches: The Riddles of Culture* (New York: Random House, 1974).

15. The classic example offered by Harris is the seemingly foolish treatment of cows in India: it makes good sense for the people in India to do what they do, but not for the (emic) reasons they themselves supply. It is for Harris, the (etic) observer and commentator, to articulate the real rationale for their treatment of cows: see *Cows, Pigs, Wars, and Witches*, pp. 11–32.

in the application of psychoanalysis to history. The result, of course, is that an agent's own narrative does not just get refined but is drastically rewritten.[16] It is possible, especially when dealing with historical figures who are dead and gone, to rewrite someone's story in psychoanalytic terms and categories: the plausibility of the new narrative thus generated depends in good measure on the plausibility of psychoanalysis itself. For the sake of argument and demonstration, I will not quarrel here with the main psychoanalytic doctrines; instead I will ask what effect they would have on our narratizing and self-consciousness if they were both true and widely accepted. That effect, it seems to me, is that they would *not* allow overriding descriptions to flourish in the long run. Psychoanalysis is itself a historical phenomenon in that the awareness of its categories and patterns in a culture changes the very situation in which psychoanalysis is practiced. It was only for a generation or two that Freud and company could deal with naive or virgin "patients" to whom to the psychoanalytic interpretation and rewriting of the dynamics of their inner life was entirely novel. Nowadays, people with a bit of education have heard the Freudian account of human life somewhere along the line and are not surprised when a psychotherapist brings it into the discussion as a framework for understanding.

The conclusion I would offer is that overriding descriptions are always possible in principle and should sometimes be regarded as legitimate. Yet they will remain the exception. The reason why they are not likely to become standard operating procedure is that there is considerable interaction between such sectors of our culture as liter-

---

16. Roy Schafer applies narrative categories in his version of psychoanalysis: see "Narration in the Psychoanalytic Dialogue," in *On Narrative,* ed. W. J. T. Mitchell (Chicago: University of Chicago Press, 1981), pp. 25–49. What happens in the process is this: "Repeatedly the analysand's stories (experiences, memories, symptoms, selves) go through a series of transformations until finally they can be retold not only as sexual and aggressive modes of action but also as defensive measures adopted (out of anxiety, guilt, shame, and depression) to disguise, displace, deemphasize, compromise, and otherwise refrain from boldly and openly taking the actions in question" (pp. 35–6). Yet Schafer allows the patient or client (analysand) a greater role in reshaping the story than earlier, more rigid psychoanalysts were inclined to do.

ature, history and psychotherapy, on the one hand, and everyday life, on the other. To the extent that something novel like psychoanalysis gets established and accepted in a culture, it works its way into the literature and other art forms (including film), and thereby into everyday consciousness. Its employment on the part of historians then tends to flow more and more along "emic" lines; that is to say, the historian does not overrule or override an agent's account of his own actions to the same dramatic extent as Freud and others did when re-writing the lives of historical figures who knew nothing whatever of psychoanalysis. Instead, people come to speak and think in terms of the (Freudian) sexual dynamics operative in their relationships with their parents; they wind up regarding such dynamics simply as part of human life, that is, as "natural" components of the human situation. And the historian, influenced by Freud, trails along behind them, agreeing with them some of the time, but also correcting them here and there.[17]

In the case of Harris I would not offer such a conclusion. When he develops his overriding descriptions of what goes on in India, or in some tribal society where there is an orgy of pig-killing every so often, he does not do so as a member of the society he is talking about. Because he is a visiting social scientist doing research, Harris speaks as an outsider, and his ideas do not penetrate or shape the society he is studying. But historians are rooted in the soil of common sense or ordinary consciousness, and usually they do not stray very far from language, categories and motivational analyses of the sort that the man on the street is able to understand. Thus their descriptions, to the extent that they are effective and revealing, will eventually merge with common sense. The result, then, is that overriding descriptions will not be the order of the day but the exception.

Non-narrative history. A good deal of work done by historians does not serve to illustrate the contribution of memory and narrative to historical truth. Not everything written by historians needs to take

17. I remind the reader that I am *not* endorsing the Freudian account of human nature here: see the previous paragraph.

the form of a narrative, and much of our interest in the past is about conditions and circumstances that do *not* lend themselves to narrative presentation.[18] For example, it is helpful to know what the climate of an area being studied was like a thousand years ago. If the climate was different, what sorts of plants and animals were common then? There may be geographical changes that relate to the changes in climate. All of these differences would have an impact on the way people lived. And that impact would in turn be felt in economic life. Moreover, historians (along with social scientists) investigate the development of tools and technology, which also influence everyday life. Archeology, as a branch of history (or perhaps a closely related discipline), tries to reconstruct past conditions of life on the basis of historical evidence (traces of the past) apart from records and sources. There are also historians who gather large quantities of data about economic and social conditions and then subject them to extensive analysis, with the help of computers.

The work of all these investigators contributes to narrative history in that it supplies us with a body of information on which we can draw when making inferences about what must have happened. The primary material with which the historian works is recollected narratives. He supplements the recollections with evidence and sometimes revises them in the light of the evidence. But as he goes about this work, he appeals not just to assumptions about human nature and behavior but also to his knowledge of the social, economic and climatic conditions of the time. Thus the conclusions of non-narrative historians do have an impact on narrative history, even if their results are not stated in narrative form.

---

18. Yet sometimes those conditions are made vivid by way of fictional narratives. Such an approach is effective especially for teaching history to children. And adults rely on it when they read historical novels, which both entertain them and give them a sense of the past. The main objection to historical novels, from the standpoint of this essay, is that when they include actual historical figures as characters, they create confusion about what really happened: the things that are said by the historical figures, for example, may be grounded in sources, but in many instances they are made up. The same problem arises with television programs in which historical events and contemporary situations are turned into "docu-dramas."

Metahistory. There are certain limits to the quest for historical truth that simply cannot be overcome. We must bear those limits in mind as we explore the sources of disagreement; in positive terms, we could say that an understanding of those limits can help us maximize agreement about the past.

The limits I am think of in this context arise especially in the constitutive element of historical thought that we would normally call presuppositions. There is a problem with this term, however, for it suggests that the person holding the presuppositions realizes what it is that he has assumed, whereas in many cases the assumption or presupposition is unconscious and is so deeply rooted that it cannot easily be stated or talked about.

Presuppositions or assumptions represent a third major component or element making up historical thought and providing material for our narratives. (The first two are primary narratives, which are the ones we start with in the process of refinement and revision, and historical evidence, which includes records and traces of the past.) One important form in which the doctrine of historical presuppositions has been developed is Hayden's White's theory of "metahistory." White maintains that the great historians have a strong leaning to one or another kind of plot structure: indeed, their loyalty to one — as opposed to others that are available — contributes to their greatness as historians. He then correlates the major types of plot structures that are used with the "four master tropes" or figures of speech.[19]

White's theory of "metahistory" represents a novel way of getting at certain questions that are often summed up under the label "speculative philosophy of history" (which is then distinguished from "analytical" philosophy of history). The "speculative" philosophers of history are the ones who develop an overall interpretative framework in terms of which they read the historical record and construct sweeping narratives. But the framework they use is not completely derived from

19. White develops and illustrates his theory in his book *Metahistory: Forms of Imagination in the Nineteenth Century* and also comments on it in some of his later writings (see Chapter 5, Note 5).

the historical material itself — neither from historical evidence nor from recollection. Therefore it has a "meta" character — paralleling *meta*physics.[20]

White's plot structures and master tropes are also to be understood as alternative approaches that cannot argue with one another on a common level in such a way as to be able to settle the issue between them. Just as some philosophers of science talk about *metaphysical* issues that arise in connection with scientific knowledge, by which they mean matters that cannot be settled empirically or by reference to scientific argumentation, so White and the speculative philosophers of history find a "meta" component in historical thought.[21]

The point that needs to be stressed in connection with this "meta" component is that there is still plenty of basis for meaningful argumentation about the past. There are indeed some "meta" factors at work, and we will probably never be able to eliminate them from historical reflection or render them neutral in terms of their effect; yet this realization should not prevent us from engaging in a critical analysis of recollections and their roots in memory.[22] And if we are aware of some of the patterns by which memory operates as it stores, compacts and edits its material, we can look for distortive factors in our own memories and make corrections. In addition we can appeal to sources and records and traces of the past, provided that we accept

---

20. Metaphysical reflection goes beyond the strictly physical and introduces considerations and categories of an entirely different order. Meta-ethics likewise operates on a different plane than ethics itself. The "meta" level of analysis and discourse is usually regarded as inherently philosophical.

21. White's position differs from the traditional speculative philosophers of history in one interesting and important way. While a "speculator" would maintain that his own grand interpretation of history is correct whereas all the other interpretations are mistaken, White seems to regard them all as having validity and plausibility, although there is an indication in his writings that the historian who grounds his approach in irony has somehow gone beyond the others (see *Metahistory*, pp. 37-8, 232–3).

22. There is a tendency in Christian academic circles to account for academic disagreements too much in terms of presuppositions and "meta" considerations, as though such considerations *dictated* all the specific conclusions we reach as scientists and historians.

them as legitimate, that is, as not having been falsified or made up. Our narrative must comport with them, and sometimes we may even have to adjust our memories to be in accord with them.

> *In aftertouching memory I try to formulate patterns; I want every gesture, every incident to fall into place as part of a becoming.*
>
> –Nina Schneider[23]

Narrative economy. Another major source of disagreements and differences[24] in historical narration comes into view when we recognize that people use a variety of strategies to respond to the need for narrative economy. (Every storyteller winds up leaving things out: we do not have enough time either to tell or to hear "the whole truth.") But here the situation is more hopeful than it was in connection with the "meta" considerations in history. For the most part, disagreements arising from the need to be succinct can be dealt with in a fruitful manner, just as disagreements over the the minutes of a meeting can often be resolved by producing a longer and more comprehensive account of the meeting.

In actual practice, however, agreement is not always easy to reach. When we economize in setting up a structure for our narrative, we are striving for a certain elegance; thus artistic considerations come into play. In principle there is no reason why aesthetic factors should enter the picture when it comes to historical accuracy and fullness, but in practice they do. We have become accustomed to artfully constructed narratives: indeed, some of the finest historians are praised for their exemplary writing style and are sometimes read for sheer enjoyment (especially after their narratives have been superseded and are judged out of date). On the level of memory, we already strive for regularization and aesthetic elegance.

The so-called synoptic problem in New Testament theology illus-

---

23. *The Woman Who Lived in a Prologue* (New York: Avon Books), 1982, p. 36.

24. I add the word "differences" because we should not assume that whenever two narratives differ from one another in speaking of the same event or set of events, they are contradicting one another.

trates what attitude we might take when we encounter narratives about the same event or set of events that differ from one another because their authors have responded in different ways to the need for narrative economy.[25] The gospels attributed to Matthew, Mark and Luke respectively differ from one another on various points:[26] yet all are accepted by orthodox, Bible-believing Christians as inspired, faithful and reliable accounts of the life of Christ. Each one has its own slant, which commentators and preachers love to explain and emphasize. No one of them claims to tell *everything* about Christ's life and ministry.[27] Christians have learned to accept all three synoptic gospels on an equal basis; in many cases — not all, of course, — something similar is possible when historical narratives seem to paint different pictures of the same event.

Finally, it is important to note that narratives never simply *mirror* events or somehow represent all aspects of the events they tell us about. There is no isomorphism, no complete parallelism in structure, between events and our narrative presentations of them. Hence all narratives are open to improvement; they can always be made more explicit. No one narrative concerning a certain event may become the standard by which all other narratives about that event are judged. Instead it should be possible to combine narratives which have the potential to supplement one another, with an eye to editing and trim-

---

25. Matthew, Mark and Luke are called the "synoptic" gospels. Since they have a good deal of material in common, the "problem" is how seeming discrepancies between them are to be accounted for.

26. Some of the differences concern the *order* in which the events making up Christ's ministry are presented; others have to do with details which are included in one gospel's account of a certain incident but are left out in another gospel's account. For example, Mark tells us about a follower of Jesus cutting off the ear of the high priest's servant when the soldiers came to the garden of Gethsemane to arrest Jesus. Luke includes this story as well and adds that Jesus healed the man. In Matthew we find that Jesus also rebuked the follower who took up the sword. And when we look beyond the synoptic gospels to John, we find that the follower in question was Peter, and that the high priest's servant was named Malchus (see Matt. 26:51–4; Mark 14:47; Luke 22:49–51; John 18:10–11).

27. John concludes his gospel by observing: "But there were also many other things which Jesus did; were every one of them to be written, I suppose that the world itself could not contain the books that would be written."

ming down the story afterward in a way that still yet includes all the details from the separate narratives that were brought together. This is in fact what happens when "story Bibles" take what we learn about Jesus' life and ministry in the four gospels and combine the material to form a single, coherent account.

Chapter 9

Contra Historicism

I believe I've sprung a leak. It's like maybe sawdust's coming out of me. Like maybe now I got to speak, on account of some-body's *got to speak — and maybe talk his way through to the truth of all this. The truth's important. It helps cause the sun to come out and the birds to sing and the clock not to fall down in a heap, you know?*
–Don Robertson[1]

What's the problem? In most cases, a word with an "ism" suffix stands for some sort of aberration or mistaken way of thinking.[2] Historicism must therefore be a diseased version of historical thought or consciousness. What is it about historical thought that has turned sour or gone off the rails?

The definitions of "historicism" we find in scholarly studies and reference works are so numerous and varied that one almost despairs of getting to the central idea.[3] A few thinkers who discuss this term or its equivalent in other languages praise historicism as a great intellectual achievement,[4] and there are definitions that include histor-

1. *Prisoners of Twilight* (New York: Crown Publishers, 1989), p. 36.

2. There are, however, a few prominent exceptions, such as altruism, evangelism, and even anachronism!

3. For a fuller characterization of historicism, especially in relation to the danger of relativism, see my book *Historical Understanding in the Thought of Wilhelm Dilthey* (Toronto, Buffalo and London: University of Toronto Press, 1980), pp. 122ff.

4. See, for example, Friedrich Meinecke's history of historicist thought, entitled *Historism: The Rise of a New Historical Outlook*, tr. J. E. Anderson (London: Rout-

icism's problematic elements without expressing any dissatisfaction with them.[5] But most characterizations suggest that it is a way of thinking that creates perplexity and gives rise to concern, and thus is something that should be "overcome" or dealt with.[6] In this latter vein I now wish to explore the question what response to historicism can be offered in the light of the conceptions of memory, narrative and historical truth I have developed above. My concern is not just with historicism as a philosophy of history but also with its (existential) implications for everyday life and attitudes.

If historicism is a disease, its first manifestation is its effect on values: we are told that it works as an acid, eating away at values and norms. For the historical consciousness that is infected with historicism, nothing is permanent and abiding; everything is in flux. There are no norms above the historical process itself. What people call morality is no more than an ethos and is always peculiar to a community: there are no universal moral standards.[7] So-called norms are rooted in specific cultural and social conditions. Hence, when you are in Rome, you ought to do as the Romans do, but their customs will not be appropriate for Paris. You have no business criticizing the

---

ledge & Kegan Paul, 1972).

5. For example, Roger Wines defines "the lasting contribution of German historicism" as ". . . the realization that all human institutions and values have their origins within the historical process; that the nature of things can be understood only in relation to this genetic process; that each individual and each institution are unique and cannot be comprehended fully in any abstract definition or principle; and, finally, that this uniqueness can be studied only through the critical use of historical sources" ("Introduction," in Leopold von Ranke, *The Secret of World History: Selected Writings on the Art and Science of History* [New York: Fordham University Press, 1981], p. 13).

6. Georg G. Iggers and Konrad von Moltke observe: "The term *Historismus* refers primarily to the historical faith of the German classical tradition of historiography. The term 'historicism' came into English in the 1930s apparently as a translation of Croce's term, *storicismo.* Since then it has acquired connotations of radical historical relativism." See their "Introduction" to Ranke's *The Theory and Practice of History* (Chapter 7, Note 16), p. xviii.

7. Kant believed in universal morality: his categorical imperative or supreme general principle of morality was valid for all human beings. But Hegel, exemplifying historicism, made morality specific to a community.

Romans when in Rome.

This way of thinking about values and morality leaves peoples and communities undisturbed[8] and deprives us of a necessary component in our confrontation with totalitarianism. Just as some people in the 1930s thought the Germans ought to be left alone to live their life in their own peculiar way (however repugnant it might be to the English, for example), there are people today who have deep misgivings about moral criticism which one nation directs at another. When a nation (or its leaders) calls for the end of certain repugnant practices (such as torture and slavery) which another nation permits, or when it appeals for a greater recognition of "human rights," it is flying in the face of such historicism. Likewise, when missionaries go out in the name of Christianity and try to wean people in faraway countries away from their native religious practices and traditions, they are also violating the historicist understanding of the autonomy of each society and cultural unit.

The other side of the historicist mentality is its view of our knowledge of the past. Just as norms and moral convictions have no validity and application beyond our own society and cultural setting, so our attempt to gain objective knowledge of people and groups in the past is bound to be frustrated, for we inevitably assimilate them to *ourselves* and force them into *our* framework of reference. We can only understand things in terms of the present, in terms of *our own* time and our needs.[9] The individual in the present is like a chip floating on the stream of history: he can gain some dim awareness of what is immediately around him, but the fuller, more objective knowledge of the past for which earlier historians pleaded is beyond him.

This does not mean that history, understood as the study of the past, is a waste of time; many historicists have been deeply committed students of the past. The point, rather, is that what we find in the past

---

8. Out the window goes the critical examination of human life, including one's own practices and commitments, which philosophy is supposed to foster and which Socrates demonstrated in ancient Athens. When in Athens . . . .

9. It should be noted that such doubts about knowledge are the consequences of *later* historicism; earlier historicism was more optimistic about the prospect of reliable knowledge of a past era.

is, to a considerable extent, what *we* have put there. The Middle Ages as described by a historian of the seventeenth century will inevitably look different from the Middle Ages as portrayed by a nineteenth-century historian. Each age produces its own history of earlier eras, and sometimes, when later generations look back upon the histories that have been produced, they find them to be almost as instructive about the eras that produced them as about the eras they purport to describe. Eventually we conclude that our knowledge of the past is largely a projection of present needs and interests.

When history is understood in such term, the project of writing history gets absorbed into the arts; it becomes a mode of fiction, a great work of human imagination. The idea of being instructed by truths about the past that rise above the tide of history and are relevant for every age of mankind is then dismissed. The elusive past can live only in the bosom of the present: the past is no more than what I make of it when I encounter relics and records of bygone days.

This romanticizing and aestheticizing of historical thought has contributed to the decline of history as a subject taught in schools. If truth in an old-fashioned sense is not attainable in history, perhaps we should relinquish the quest for it and seek (educational) solace elsewhere, e.g. in the social sciences. Thus the effort to respond to historicism is bound up closely with the question of the role of history in general education. Christian education has traditionally insisted on a very important role for historical study; this is one of the differences between Christian and public schools today.[10]

Presentism. The preoccupation with the present that is characteristic of historicism is built into the very methodology of historical study as understood by historicist thinkers. For the nineteenth-century German historian J. G. Droysen, the past was not simply to be resurrected; rather, in the words of Hayden White, the historian was to aim for an "unveiling and exaltation of the present that was latent within

10. Sad to say, some Christian day schools, under the influence of public and secular models in education, have let history slip in terms of the prominence they assign it in their curriculum.

it," and, conversely, he was to "enrich the present through explication and illumination of the past that inheres within it." White speaks of this approach as "presentism,"[11] for it turns the present into a filter and framework for our knowledge of the past. We might ask in response: does the past not make its own demands on us? Are we not able to be challenged and surprised by the past? When we think historically, do our present circumstances limit and constrain us to so great an extent as "presentism" suggests? Those who have succumbed to historicism would answer this series of questions with a yes.

In the hermeneutics of H. G. Gadamer we are offered a partial answer to this presentist bias within historicism. Gadamer, as a lover of the classical world, assigns a certain power and independence to classic texts such that their testimony to us is not simply a function of how we, with our present needs and concerns, are able to view and read them. His notion of the "fusion of horizons" is an effort to break through the confines of historicism by assigning to the past and to our encounter with it a substantial power and influence in our cultural and intellectual life. And if we are to seek a *fusion* with the past, the world of the text must possess some substance and contours of its own; our standpoint in the present does not serve to determine what the past can mean to us or to set limits to its meaning.[12]

We can also point to the role of retrospective and prospective discourse in human life as we seek to counter the tyranny of the razor-thin present and thereby break with presentism. At first glance, of course, the notion of retrospective discourse seems to play into the hands of historicists, for it appears to suggest that we constantly remake the past in the language and imagery of the present. But when we look at prospective and retrospective discourse *together,* we see that using both enables us to live in an enlarged present in which the past (or at least some substantial portion of it) is with us constantly as a part of our life. The temporal priority of the past over the present means that it is the past that has, in good measure, made the present

11. See White's essay "Droysen's *Historik:* Historical Writing as a Bourgeois Science," in *The Content of the Form* (Chapter 5, Note 5), pp. 94, 91.

12. See *Truth and Method* (New York: Crossroad, 1984).

what it is today. Therefore it would be a mistake to suppose that we have *revived* the past by our activity as historical thinkers and writers, breathing life into something which, apart from our efforts, would be no more. Instead we should recognize that the "elusive past" never was very far from us: so much is it a part of our life that we update and adapt it constantly, not just in the light of present needs and concerns but also with an eye to the future, to our intentions and projects (prospective discourse).

The best answer to historicist presentism, then, is to affirm that we do *not* live in a razor-thin present in which the past can exist only by virtue of our historical consciousness; instead we should speak the covenantal language of the Bible, which presupposes the continuing reality of the past. The Biblical believer lived in an expanded present in which God's deeds and promises in the past formed the foundation of his existence. In the last hundred years or so, Christians who live by the Bible have had to expand their sense of the temporal framework of human existence, just as they have come to recognize that their world is physically larger, in both an abstract and a concrete sense,[13] than earlier people had imagined. To live covenantally — in partnership with God — is to "think big." Therefore the ideal locale for Christian worship is a *big* sanctuary, preferably with a roof that is much higher above the heads of the worshippers than the roof in their living room at home.[14] "How great thou art!" we sing. By being linked to God, for whom a thousand years is as a day (Psalm 90:4), we also take on an expanded sense of time and place. The kingdom of God is without boundaries.

This recognition of the expansiveness of the present should bear fruit in our Bible reading. There are some Christians who fall prey to

---

13. The world is concretely larger in that we range over much more territory than people used to and are able to move from place to place remarkably quickly; we are no longer confined to our village but can live our lives in different locations, sometimes on separate continents, between which we can move back and forth easily. Just as the world of a child, i.e. his home town, seems small when he later looks at it through adult eyes, so the world inhabited by a medieval peasant would seem small to a twentieth-century person.

14. Some shopping centers are architectural rivals to churches in this respect.

historicism by allowing their own experience and needs to serve as a filter through which the Biblical message must pass. ("What does this text mean to *me, today?*") Something of this mentality is likewise present in Christians whose eschatology we might describe as overheated. From a presentist standpoint, the Biblical passages about the last things must have application to *my* life in *this* period of history — or else they are meaningless. In positing such an either/or and then shying away from the latter conclusion, such "presentist" Christians are driven to overinterpret the eschatological passages and force them into a mold which allows for point-by-point applications to today's events and which can only result in sensational conclusions. The people who read the Bible in this fashion need to expand the temporal framework within which they live and think.

> *The only standards that remained were of a purely subjective character, standards that had no other support than the free choice of the individual. . . . Historicism culminated in nihilism.*
>
> –Leo Strauss[15]

Egocentrism and ethnocentrism. If historicism is guilty of not thinking in big enough terms, the problem lies not just in its conception of time but also in its preoccupation with a subject who is too limited. For existentialists, who can be understood as radical historicists substituting a separate (alienated) individual for the people or "Volk," everything must relate to *me* in order to be real and valid and have meaning. Original historicism had instead made the national community the focus of all meaning. In its recognition of a corporate dimension in human experience, such historicism does have something in common with the Bible, especially the Old Testament. But we must also consider what the New Testament says about the people of God.

Historicism can be understood as a mild version of what philosophers call solipsism, which is a belief in the self as the sole reality (just me myself, according to the Latin etymology). In its strongest form, solipsism is the belief that only I exist: everyone and everything

---

15. *Natural Right and History* (Chicago: University of Chicago Press, 1953), p. 18.

else is a figment of my imagination.[16] While such an extreme solipsism is not usually propagated by philosophers (since there would be no one to teach it to), a weaker form of solipsism is fairly widespread. It maintains that relevance must be understood in terms of the individual: something can be important only if it relates to *me.*

This is of course the existential posture adopted by an immature child, a posture normally outgrown at some point in childhood. Yet a preoccupation with oneself as the sole reality or the only person to be taken into account can also be found in certain figures in the adult world. It has the effect of reducing the self to a series a fleeting experiences having little connection with one another. Such a self cannot very well have relationships with other people either. Oscar Wilde says of his famous character Dorian Gray: "He used to wonder at the shallow psychology of those who conceive the Ego in man as a thing simple, permanent, reliable, and of one essence. To him man was a being with myriad lives and myriad sensations . . . ."[17]

It is interesting to note that one dimension of the normal process of maturation is in effect an overcoming of historicism: what happens — or *should* happen — as we grow older is that we get beyond an existential solipsism and egocentrism as we identify with a wider human community and come to recognize the validity of its aspirations. The ability to identify with a wider community and with people who lived long, long ago is also essential to developing an existential sense of what the doctrine of original sin means. What does it mean to be implicated in the sin of Adam and Eve? If we think along historicist lines, this doctrine will remain a mystery to us.

Historicism expands this solipsism and extends it to a group or "Volk" or nation. Once we recognize these solipsist roots, we see that historicism does have some relation to the challenges of fascism and racism we have confronted in the twentieth century: the inability to recognize the rightful place and full humanity of groups and persons

---

16. Solipsism in this form is explored by Descartes in his *Meditations on First Philosophy:* see especially the first meditation.

17. *The Picture of Dorian Gray* (London: Unicorn Press, 1945, originally published in 1891), p. 187.

that are racially and culturally different from one's own is bound up with historicist short-sightedness and lack of expansive thinking. (Of course I do not mean to deny that racism also has some roots or sources that have nothing to do with historicism.)

The Bible represents an antidote to historicism. The theme of the stranger in the Old Testament (e.g. the stranger was to share in the rest of the sabbath day) is an indication that God's revelation to his people long before the time of Christ was intended to lead them away from any ethnocentric solipsism or preoccupation with themselves. God had said to Abraham: "By you all the families of the earth shall bless themselves" (Genesis 12:3). And the New Testament recognition that the people of God are drawn from every nation and are not confined to any one geographical territory should enable Christians to grow beyond the self-centeredness that asks, in every situation, what this has to do with *me* (as an individual) or perhaps with *my group* or *my tribe.*

Salvation in Christ makes the believer part of a body that spans all times and peoples. He learns to identify with earlier people as having undergone a history that is part of his own history. Not only the ancient Israelites but also Christians today must learn to say of Jacob: "A wandering Aramaean was my father; and he went down into Egypt and sojourned there, few in number; and there he became a nation, great mighty, and populous. And the Egyptians treated *us* harshly, and afflicted *us,* and laid upon *us* hard bondage. Then *we* cried to the LORD the God of our fathers, and the LORD heard *our* voice . . ." (Deuteronomy 26:5–7, italics mine –TP).

It is not good that the man should be alone.
Genesis 2:18

Marriage. The institution of marriage can also be regarded as an existential antidote to historicism. As Kierkegaard, in particular, has shown us,[18] being married requires of us that we stand in a certain

18. See "Aesthetic Validity of Marriage," which is part of *Either/Or,* tr. Walter Lowrie, with revisions by Howard A. Johnson (Princeton: Princeton University Press, 1959), pp. 3–157 of Volume II. It should be borne in mind that the thinking of the

relationship with time that breaks through the limitations of the razor-thin present; in fact, the appeal which marriage clearly holds for many people is inconceivable apart from this experiential possibility. Marriage represents shared experience over a long stretch of time; it makes possible an experience which issues from a sexual union and which (under normal circumstances) produces children who can be enjoyed in a way which is thoroughly temporal. Moreover, the coming of children binds the present to both the past (through precious memories) and the future (the relationship between parents and children is lifelong, even though it changes as the children mature). Through marriage and the birth and raising of children, one's life takes on an existential fullness that plays a major role in the experience of time, with the successive instants of which life is composed (when abstractly conceived) coming together to form a coherent whole that helps to anchor one's selfhood.

Kierkegaard speaks of a "unity of hope and recollection" which he identifies with "the true present," warning that we neither "predominantly live in hope" nor "predominantly live in recollection." Marriage represents life in a full present that includes a continuous awareness of its own past (the "historical factor") and its hopes for the future. Through marriage we are able to "contend with time." Of the married man (who is contrasted with the aesthete, living only in the present), Kierkegaard writes: "He has not fought with lions and ogres, but with the most dangerous enemy: with time." Yet this ideal husband cannot be represented, for artistic representation focuses on the momentary: "Romantic love can very well be represented in the moment, but conjugal love cannot, because an ideal husband is not one who is such once in his life but one who every day is such."[19]

Here Kierkegaard points out a reason why the existential possibilities with which marriage presents us are not widely known and ap-

---

individuals who "speak" in this work cannot simply be equated with Kierkegaard's own convictions: *Either/Or* is to be read as a dialogue between spokesmen for "the aesthetic" and "the ethical." The explanation and defense of marriage is undertaken by the proponent of "the ethical."

19. The quotations and references in this paragraph all come from pp. 138–46 of "Aesthetic Validity."

preciated: the Romantic (aesthetic) preoccupation with the momentary lends itself much more to artistic glorification than does conjugal love. Kierkegaard observes: "You may convince yourself of this more thoroughly by considering the predicates commonly applied to conjugal love. It is faithful, constant, humble, patient, long-suffering, indulgent, sincere, contented, vigilant, willing, joyful. All these virtues have the characteristic that they are inward qualifications of the individual. The individual is not fighting with external foes but fights with himself, fights out of love from within him. And they have reference to time, for their truth does not consist in being once for all but in being constantly what they are."[20]

Kierkegaard discloses something of what marriage makes possible by way of his brilliant depiction of the "seducer," the Don Juan figure.[21] The point I would add and stress is that the ability and opportunity to experience what Kierkegaard's "seducer" has missed depends in part on social, cultural and legal decisions, especially in relation to marriage and family. On this collective level, too, there is a concrete choice to be made for or against historicism. In a society determined to combat the historicist mentality, there will be plenty of legal and social space for long-term relationships and life-patterns, of which marriage is the most prominent.

Don't talk about horrid subjects. If one doesn't talk about a thing, it has never happened.
–Oscar Wilde[22]

Cynicism and denial. The historicist mentality, with its denial that there are any values and norms enjoying abiding validity and its skep-

---

20. "Aesthetic Validity," p. 142. Although marriage is very often depicted in literature and the modern novel, what we actually read about is *conflict* within marriage. Ideal marriage is usually depicted only at the beginning of such a work (before the breakdown occurs), or at the end (in case the conflict is overcome and the marriage relationship is fully restored). The marriage relationship is not easily depicted apart from some sort of struggle or conflict.

21. See "Diary of the Seducer," which makes up pp. 297–440 of Volume 1 of *Either/Or*.

22. *The Picture of Dorian Gray*, p. 141.

ticism regarding knowledge of the past, has much cynicism in the last century or so. We see such cynicism in the reaction of many people to the famous Nuremberg trials held shortly after World War II, at which Nazi war criminals were brought to justice through judicial proceedings run by the victorious Allied countries. Cynics complained that what happened at Nuremberg was just a more ornate version of what has happened after a great many wars, namely, that the victors punished and humiliated the vanquished, which does not do much to ensure a genuine and lasting peace, since the resentment it creates can easily fuel a new armed conflict. This would seem to suggest that the real crime of the Nazis who were tried was that they lost the war. While most sane and sensible people would recognize that there was a great deal more wrong with the Nazis than their military miscalculations, the tendency to regard the Nuremberg trials as a farce does have some support in the West. And that support is fueled by historicism.

The cynics say that after an armed struggle, it is the privilege of the victors to destroy the evidence of their own wrongdoing and then to deny their misdeeds altogether. What is no longer spoken of does not come readily to mind and is eventually forgotten. And so history is written by the victors. The "moral of the story" is that they were in the right. Indeed, what could possibly be meant by "the right" other than the cause of those who triumphed? Thus tourists today have opportunity to behold concrete reminders of Hitler's crimes against humanity, for he was not around at the end of the war to remove them, but Stalin's misdeeds are harder to trace and remember, for he was never overthrown and thus had the opportunity to hide much of the evidence, thereby influencing history's account of his regime.

Unfortunately, the cynics who take this line are not altogether wrong: there is probably a good deal of horror perpetrated by Stalin and his followers that by now is known only to God. Yet when we seek to overcome the excesses of historicism, we may certainly take heart from the determined efforts which both survivors and historians have made to bring out the story of Stalin's brutalities. Despair about truth in history is not justified when we take the case of Stalin as an example.

An important element in his case and in the outrage his misdeeds have unleashed is that some of what we know comes to us in the mode of confession. Certain of the evildoers formerly enrolled under his banner saw the error of their ways and told us what was going on, usually after escaping to the West. Of course their accounts of their own involvement cannot help but be somewhat self-serving, but we should not rule out the possibility that such people are possessed of a genuine desire to have the truth come out.

Confession — that purging of the soul which is emphasized and promoted especially by the Christian tradition — is a powerful weapon against historicism. It shows us that there is a route open to us other than denial. We can deal with what we have done wrong by confessing it, seeking forgiveness, making restitution if possible, and then putting it all behind us, for with forgiveness comes renewal and the possibility of starting over on a new footing. Cynicism and denial may not have the last word when we contemplate the past and our role in it.

Genesis and validity. Another manifestation of the historicist outlook is its tendency to settle the question of the validity of a social arrangement by inquiring into its genesis or origin. Now, at first such an approach does not seem objectionable or strange; indeed, it appears to be a species of conservatism,[23] and it does translate into support of the existing order. But such support does not always meet the demands of justice. In this area, too, the historicist approach must be challenged.

The first danger to be noted is that the preoccupation with origin as establishing validity can lead to the temptation to confer a false pedigree upon oneself or one's group. If those who occupied a piece of ground first are, by virtue of this fact, its rightful owners and therefore are entitled to control it and to receive the lion's share of any revenues that might be derived from its crops and natural resources,

---

23. While some early historicists are regarded as conservatives in their political thinking, we should not equate conservatism with historicism, for conservatism also has roots in "natural rights" thinking. On this question, see Leo Strauss, *Natural Right and History.*

the battle or legal contest over possession will presumably be won by whoever comes up with the most impressive evidence of early occupancy. The quest for such evidence can lead to historical falsification of a kind that is considerably more serious than the efforts people sometimes make to establish themselves as being of noble or even royal descent. Yet in communist countries, pedigrees were often falsified in the opposite direction: it was politically expedient *not* to be descended from the aristocracy but to claim roots in the peasant class instead. And then there are cases in which the prize is "victim" status: one must show that one's forebears were persecuted, mistreated, discriminated against, in which case it becomes possible to claim "protected minority status." All such efforts to root one's recognized needs and rights in historical conditions tend to promote dishonesty.

The second reason why the historicist link between genesis and validity must be resisted in that it may lead to excessive rigidity in our thinking about social and ethical matters. It must be possible for us to recognize *new* needs and *new* situations developing. In response to new challenges, we must not rest content with appealing to precedents and guarantees made to earlier groups: instead it becomes necessary to think in terms of the demands of justice in a more general sense. New and unprecedented damage to people's property, livelihood and health because of an environmental disaster (e.g. a major oil spill or a nuclear accident) will require imaginative solutions and applications of justice.

The world we live in, including the set of conditions under which we live, changes in some fundamental ways over time, ways which could not have been anticipated by earlier generations, and when it does so, we must be willing to develop new ways of getting along with one another and respecting each other's prerogatives and needs. The historicist preoccupation with genesis and precedent will not serve us well when we need to think about property rights in the light of a changing climate that may render formerly habitable parts of the earth barren and may convert land into sea.

Upheavals in history, such as the displacement of entire populations caused by wars and atrocities, also require that the human community at large confront *new* needs that are not sanctioned by precedents and long-recognized arrangements. The Palestinian conflict, for example, will not be solved as long as people persist in thinking that there must be a historically "correct" answer to the question "Whose land is it?" Refugees cannot be treated as trespassers and told to go back home. When we ask whose responsibility it is to take them in and give them a home, there is no answer forthcoming from the domain of historically sanctioned rights and prerogatives.[24] It seems that all the nations could get by with saying no; the result is that the entire world then has a festering sore to contend with and may well have to pay a heavy price for its failure to promote peace and justice.

Likewise, it is important to recognize that some conflicts have gone on for so long and have become so bitter that there is no way to sort them out in terms of right and wrong. The conflict in Northern Ireland, for example, is not one that can be neatly analyzed by lawyers. In the final analysis one must hope and pray for an attitude of mutual acceptance (an application of what Christianity calls "loving one's neighbor") that finally recognizes the enemy as a fellow member of the human community and extends the olive branch of peace to him and allows him a home on the earth. In earlier centuries, armed quarrels sometimes led to the extermination of the entire defeated population (just as a new king might feel compelled to kill all other claimants to the throne), but in a time when technology makes it very difficult to carry out such brutal solutions while escaping the scrutiny of the world at large, it is more important than ever for warring groups to come to terms with one another, and ultimately to make peace in an attitude of generosity that is rooted not in historicism but in a Christian spirit of charity. Historicism simply does not meet the needs of our time.

Do you realize that the past, starting from yesterday, has been actually abolished? If it

24. In Biblical terms, one could argue the same point through the parable of the Good Samaritan (Luke 10:25–37), who did not look for some form of identity card spelling out the sufferer's rights or legal claim to assistance: he simply tended to the suffering man's wounds and saw to it that his needs were met.

survives anywhere, it's in a few solid objects with no words attached to them . . . .
–George Orwell[25]

Monuments and records. What is required of us if we are to overcome historicism is that we refrain from trying to assimilate the past to the present; instead we are to respect it and let ourselves be instructed by it — and perhaps even judged.[26] When we remember the story of the sacrificial death of Christ as we celebrate communion, we are reminded that those who hailed him on Palm Sunday cried out for his crucifixion only a few days later. His broken body is a finger of reproach pointing to our natural human enmity against God, which can be overcome only through the grace of God.

For the Christian, remembering is not limited to communion. Nicholas Wolterstorff speaks of remembering as "one of the profoundest features of the Christian and Jewish way of being-in-the-world and being-in-history."[27] Our hymns and psalms contain many appeals to remembrance, and in the Old Testament we come upon the notion that there must be "monuments," physical tokens of remembrance for a people among whom the ability to read was rare. For example, twelve large stones were set up as a monument at Gilgal, near the Jordan River, to remind the Israelites of how their God had kept his promise and led them into the land (Joshua 4:19–24). And when the city of Jericho fell to the Israelites, God decreed that its walls were not to be rebuilt: the ruined walls would be an abiding monument reminding his people that the city had been taken through the power of their God — and not through their own might (Joshua 6:26–7).

The Roman Catholic tradition makes more use of "monuments" in its worship traditions and in the outfitting of its churches than Protestantism does. The Protestant ethos is rooted in an era when the printing press came into use, which meant that more emphasis could be placed on reading and on the vernacular languages; hence Protes-

---

25. *Nineteen Eighty-Four: A Novel* (London: Secker & Warburg, 1949), p. 156.

26. In adding that the past may have to judge us, I am taking issue with Nietzsche, whose response to historicism (see Chapter 1 above) went too far.

27. *Lament for a Son* (Grand Rapids: Eerdmans, 1987), p. 28.

tantism is more oriented to reading, written records and "the Word." This difference between two major streams of Christianity need not detain us here. The spirit that wishes to resist historicism will want to preserve both monuments and written records — in the name of truth and historical integrity.

Many monuments testify to good and great deeds, depicting heroes on horseback, but there are also some that remind us of unspeakably foul deeds performed by man. Such, for example, are the sites of former concentration camps and death camps in Europe. They are preserved so that we will never forget what man is capable of doing to man. There is also quite a documentary record of this despicable chapter in human history.[28] Nazi photographers took pictures of many of the atrocities as they were committed and even made horrifying films of their beastly activities. Such documentary records are preserved, although some are so ghastly and gruesome that they are rarely looked at and are generally not regarded as appropriate for use in television broadcasts.[29] If we are not to let those sufferers slip into oblivion, we must look through the records and never forget what happened.

We possess those records of Nazi atrocities mainly because Nazism was defeated and the Nazi strongholds fell into our hands. When it comes to other totalitarian regimes, the records are harder to come by. Thus there has been more debate and disagreement about the nature and extent of Stalin's misdeeds, for example, than about Hitler's. The differences between totalitarian and free societies do not only concern politics and economic organization; they also have to do

---

28. For example, a remarkable book called *The Stroop Report* (published in an English edition by Pantheon Books of New York in 1979) is made up largely of photographs and documents which the Nazis themselves kept as a record of their liquidation of the Warsaw ghetto in 1943. Although the book contains an introduction and some maps and notes at the end, the documents themselves are presented (in the original German and in English translation) without comment. The book takes its title from the name of the Nazi general in charge of this operation.

29. On the occasions when I have seen some particularly ghastly film shot by the Nazis themselves, it was usually broadcast by a television station with comparatively few viewers (not one affiliated with a major network). Not many people care to watch programs featuring such film.

with respect for the past as it lives on and is available to us in written records and other traces. In his novel *Nineteen Eighty-Four,* George Orwell has given us a frightening picture of a totalitarian society that continuously falsifies records and recreates the past along lines dictated by the current ideology and political strategy — all of this in an effort to manipulate the consciousness of its own citizens. While nothing quite so effective has ever been accomplished in the real world,[30] his novel is an abiding warning that we must instill an appreciation for historical integrity and a respect for sources and records and a love for truth in each new generation. Lying and falsification may sometimes appear to be the way out of a tight corner, but in the long run we are well served by the truth.

May the LORD judge between me and you . . .
I Samuel 24:12

Judgment and vindication. People involved in conflict yearn to be vindicated, proven right, given their due. But in many cases their hopes are not realized — especially not in the short run. The historicist mentality, with its overestimation of the present, can only demand vindication in *the present,* for only the (thin) present is real. Any judgment by a sovereign God would lie beyond time and would therefore fall outside the historicist's framework of awareness.

Vindication is also a theme in the Bible, but the cry that goes up from beleaguered believers is geared to the righteousness of *God;* it is not first of all a species of *self*-righteousness. Job cries out repeatedly to be vindicated, and David, in a dramatic encounter with his enemy Saul, gives us a concrete picture of what a Biblical alternative to a historicist mentality would stress. Saul has been pursuing David, but at En-gedi David gets the chance to kill him. He refuses to do so and cuts off part of his robe instead. Then he says to Saul: "By the fact

30. Czeslaw Milosz writes: "Never before has there been such enslavement through consciousness as in the twentieth century" *(The Captive Mind* [Chapter 3, Note 2] p. 191). Perhaps the events in eastern Europe at the end of the 1980s will lead us to conclude that it is harder to take the human mind captive than some of the totalitarians had assumed.

that I cut off the skirt of your robe, and did not kill you, you may know and see that there is no wrong or treason in my hands. I have not sinned against you though you hunt my life to take it. May the LORD judge between me and you, may the LORD avenge me upon you; but my hand shall not be against you. . . . May the LORD therefore be judge and give sentence between me and you, and see to it, and plead my cause, and deliver me from your hand" (I Samuel 24:11-12, 15).

Here David appeals to circumstances and evidence in support of his interpretation of what has happened, for he points to the opportunity he has passed up and even clutches a part of Saul's garment as proof that he was within striking range. But he leaves the judgment of the entire matter in God's hands: let the righteous Judge determine the outcome! At a second such encounter on the hill of Hachilah, where he again spares Saul's life, David raises the possibility that it is God himself who is stirring Saul up to pursue him. Again he recognizes God's sovereignty as Judge. His parting words to Saul are: "Behold, as your life was precious this day in my sight, so may my life be precious in the sight of the LORD, and may he deliver me out of all tribulation" (I Samuel 26:24).

David does not seek vindication through military might and the sword. And he sends out a signal that the judgment rendered by the Lord, the vindication in the sight of heaven, might not be evident in this life. Here we are reminded that the Biblical believer lives not just in human time but in eternity; in his longing for vindication he looks far into the future. This is what Abraham must have done when God sent him on a life's journey away from the land of his fathers, and surely Noah must have been looking into the future when he embarked on the enormous project of building an ark which, as far as human beings could determine, would be useless when completed.

When we consider what it means to live in anticipation of such a vindication, one which jeering neighbors may never see and acknowledge, we understand what it is to "live by faith." The famous characterization of faith in Hebrews 11 can thus be understood as a repudiation of the historicist mentality: "Now faith is the assurance of things hoped for, the conviction of things not seen" (vs. 1). The heroes of faith we meet in this chapter had their hearts set on things beyond their own lifetime: Abraham, for example, "looked forward to the city which has foundations, whose builder and maker is God" (vs. 10). And Joseph, "at the end of his life, made mention of the exodus of the Israelites" (vs. 22), which was still centuries in the future. The fulfillment of *God's plan* — not the realization of their own hopes — was the vindication these believers sought. "These all died in faith, not having received what was promised, but having seen and greeted it from afar, and having acknowledged that they were strangers and exiles on the earth" (vs. 13). They were willing to be out of step with their own time and were content not to be vindicated in this life, in the sight of the scoffers.

Chapter 10

Conclusion:

Redeeming the Past

With television we vault ourselves into a continuous, incoherent present.
–Neil Postman[1]

Redescribing the past. The kinds of questions I have discussed in this essay are not readily understood by the average person of our time. Because our culture has led us further and further in the direction of a preoccupation with *the present*, historical awareness is in deep decline. The widespread lack of a feeling of responsibility toward one's own past, the kind of responsibility that would that might lead a person to think in terms of "redeeming" the past, is part of the reason why marriage and family relationships do not seem to flourish as they once did. The past no longer binds us. David Viscott expresses the sentiments that live in the hearts of many people today when he claims: "Life should be lived in the present, for it is only in the present that we are able to exert any control over our lives. We cannot change the past . . . ."[2]

One of the possibilities I wish to defend in this essay is that of

1. *Amusing Ourselves to Death: Public Discourse in the Age of Show Business* (New York: Penguin Books, 1986), p. 137.

2. It is significant that Viscott, a psychiatrist, makes this statement in a book entitled *The Language of Feelings* (New York: Pocket Books, 1977), p. 157. Psychiatry, under Freud's influence, has drifted away from historical truth into a preoccupation with feeling and imagination (see the discussion in Chapter 1 above).

redescribing the past in a way that could count as *redeeming* it. This possibility is rooted in the acceptance of anachronism, which, by now, should be seen as much more than "historical shorthand";[3] anachronism is part of the very process of redemption understood in its traditional theological sense. As God carries through his plan of salvation, he brings good out of bad and eventually achieves a state of things ("I make all things new" – Revelation 21:5) that calls for a redescription of the sorry story of human sin and rebellion, a redescription which (especially for Karl Barth) brings Christ, the Redeemer, into the discussion from the very outset.[4]

When Christians bear witness to one another about how God has been active in their lives, they are in effect offering anachronistic redescriptions with a redemptive theme. Consider the following narrative as an example.

> Although I was raised in a Christian home, I wandered away from the Lord when I was a teenager. He kept tugging at me, but serving him was about the last thing I had in mind. Well, finally he put me in the hospital. You see, I'd been in a motorcycle accident — actually, it was my fault, really — and I lay in bed in traction for almost eight months. For the first five months or so I was bitter. Then the sun began to shine just a little in my life again. God sent me a wonderful Christian nurse, whose testimony and example began to remind me of things I had learned at home and forgotten. And now I'm a nurse myself. If, just once in a while, I can do for my patients what that wonderful nurse did for me while I was imprisoned in traction, the misery I suffered will be well worth it.

---

3. The term "sunset" is not scientifically accurate but is used as a piece of shorthand: the person who employs it could presumably give a fuller and more adequate account of what actually happens. I maintain that anachronism is *not* simply shorthand: in other words, we should not assume that an anachronistic historical statement can be restated at greater length in such a way that there is no violation of temporal sequence.

4. Part of the reason for insisting that the Bible is not a "textbook" of history and science and so forth is that redemption (anachronistically) pervades its pages. When teachers in Christian schools offer instruction in subjects like physics and geography that seem mundane (i.e. having to do with this world), they do not appeal to the Bible in its main redemptive thrust.

In this narrative, human events are ascribed to God: he put the narrator in the hospital (although the accident was her own fault), after tugging at her unsuccessfully in her rebellious teenage years. The narrator now looks back on those events, accepts them as God's doing, and affirms them (it was "well worth it"). Thereby she claims and embraces her own past, instead of saying, as a secular person might, "What a horrible time that was! I'm doing my best to forget all about it. I don't even want to talk about it." She uses her past to tell others about her faith and Christian commitment.

Unless the LORD builds the house,
those who build it labor in vain.
Psalm 127:1

Divine involvement. Christians believe that God plays a role in their lives, but most of the time they are not sure what he is doing. Sometimes they are able to see his "hand" only after the fact. Some of them talk very freely about divine involvement in their lives — perhaps too freely — whereas others are too reticent, too afraid to draw conclusions on this matter. Consider the following story of Larry and Sally.

■ Larry had long wanted to live in southern California but his wife Sally was dead set against it. She wanted to stay in Pennsylvania, where she had grown up and lived all of her life. She couldn't bear the thought of moving so far away from her family and friends.

Larry was determined, and without telling Sally what he was up to, he often inquired about jobs in southern California. Twice he went there for an interview, without letting her know. He claimed he was making ordinary business trips. The second time he came back with an attractive job offer. But how was he to broach the subject with his wife?

"Honey," he announced, "the Lord has opened a door for us! I just got an wonderful job offer in California. What do you say?"

She was not enthusiastic. "Could it be that the devil is the one who has just opened that door for you?" she asked.

As we consider this story, we are inclined to fault Larry for not dealing with Sally in good faith: he lies to her about the reason for his trips to California. And when he adds God's approval to his plans, he compounds his deceit, and his wife counters him with a stinging rebuke. Yet we would not say that her suggestion is altogether inappropriate in theological respects: the devil does open doors for us to walk through, and so we pray, "Lead us not into temptation."

When we talk about redeeming the past and take proper account of God's involvement in our everyday life, we run the risk of misusing his name and making him the silent backer of all our plans, like a wife who doesn't say much but doesn't dissent either when her overbearing husband appeals to her for confirmation. To avoid this danger,[5] we must remember that God does not advertise his intentions to us. In Biblical times he sometimes announced via a prophet that he was "in" or "behind" this or that chain of events,[6] but we who live after the era of Biblical revelation must draw our own conclusions — cautiously, and in the light of Biblical teaching — about his involvement in our hopes and plans and deeds. One avenue we use to reach those conclusions is discussion with others. (Larry should have discussed his plans with Sally, and not just confronted her with his job offer as a *fait accompli.*)

Christian institutions are especially subject to these dangers. God does not run them in any operative sense — boards and executives and administrators do. God does not make mistakes, but the people who run Christian institutions certainly do. We ask God in prayer to bless the work of Christian institutions; yet sometimes he withholds his blessing.

When we recognize divine involvement in our lives and incorpor-

---

5. I speak of it as a danger because it makes us smug — and liable to forget that our plans and hopes are often shot through with sin and greed.

6. See Jeremiah 21, where the Lord announces he will *not* deliver his people from the Babylonian king who is attacking Judah but says instead: "I myself will fight against you with outstretched hand and strong arm, in anger, and in fury, and in great wrath" (vs. 5), and also Jeremiah 42, where the prophet tells the people on God's behalf that they must not flee to Egypt to escape the Babylonians. See also I Kings 12:24, where the Israelites learn that the impending split between the two kingdoms that came to be called Israel and Judah was divinely ordained: "You shall not go up or fight against your kinsmen the people of Israel. Return every man to his home, for this thing is from me."

ate it in our narratives about our own past, we have even more reason to avoid any claim that we "live" stories and simply tell what we have lived (see the discussion in Chapter 5 above). Someone thinking within a strictly secular framework *might* argue that I am the shaper of my own destiny, that I act freely and know what I am doing and can therefore maintain in memory a narrative-like record of what I have done, which I draw on when I am asked to tell the story of my life. But a Christian understanding of human life emphasizes that *God is sovereign,* and that his will for us includes many a surprise along the way, like the surprise that awaited Sarah when she was an old woman well beyond child-bearing years (see Hebrews 11:11). Thus we do have some sense of what is happening to us in life's flow, but if we are to tell the story properly, and especially if we propose to comment on what God was doing in and through the events in our lives (such as a motorcycle accident, or a job offer in California), we must look back and reconstruct things in the light of how they came out.

As for our efforts in the present, we need not despair of God's involvement. The Christian may certainly think of himself as God's partner in the great range of activities that fall within the framework of God's kingdom and rule, but he has no automatic assurance in advance that God will see fit to bless everything he undertakes. He knows, indeed, that much of what we try to accomplish comes to nought and is swept away. And so he prays, "Establish thou the work of our hands" (Psalm 90:17).

Condemning the past. To redeem the past requires that we *take action* at some point — it is not enough to *talk* about the past, however important talk may be in itself. We sometimes say that talk is cheap, and in many situations it is; yet there are also historical circumstances in which talk is forbidden or drastically restricted. More specifically, honest talk about the past may be forbidden. There may be things in the past that cry out to be confronted and condemned, but some tyrant or oppressor tries to silence us. (The Holocaust would not have been possible if so many otherwise decent people had not chosen to look the other way and remain silent when the mistreatment and harassment of Jews was encouraged and promoted by the Nazis well before the program of systematic killing was begun.) And so there are occasions when we must summon the courage to condemn what we see happening around us.

This project of condemning should be regarded not as the *opposite* of redeeming the past but as its *counterpart.* To make something of the past is to choose, select. We do not take all of it with us or find every part of it equally significant and fruitful. There are some parts that we must condemn and turn away from resolutely. But it is not enough to just "forget about it": what is required of us is honesty and a commitment to historical truth. We manifest that commitment first of all by refusing to falsify the past or to cover up ugly episodes or to downplay its evil, and secondly by making sure we do not forget it. And if we are determined not to forget, we must be willing to be reminded on occasion, to hear the ugly story again, to look once more at distressing photographs and film.

Sometimes we must go still further and lend a voice to those who have lost theirs. One might respond that this is the task we assign to lawyers; yet there are situations in which it is to late to do justice for the one who has been abused and mistreated — perhaps because he is dead. Such a case cannot be closed. Even though it is too late for the mistreated individual, the story still needs to get out and be brought to the attention of a wider human community. To recognize this necessity is to accept that there is a solidarity created by suffering and injustice: the sufferings of our fellow human beings (and not just of members of our own small community) must be *our* concern as well. Therefore apartheid is our business in North America.

One way to help make those sufferings and injustices known is by lending a voice. Journalists have a responsibility here, and so do writers of fictional literature; indeed, fiction is often more effective for creating a vivid and abiding sense of what has happened than is documentary history.[7] And the rest of us, who do not have the talent or

7. The writings of Alexander Solzhenitsyn have shown us something of the range of possibilities when it comes to lending an effective voice to those who can no longer cry out on their own behalf. There is important truth in the fiction he has written about Stalinist oppression.

occasion to lend a *voice* to the oppressed, should surely be willing to lend an *ear*, by hearing and reading their stories and seeing to it that those stories are passed on to succeeding generations, accompanied with the lesson that the real enemy is not this or that ethnic group but the impulse toward evil in man as such. The impulse to control, abuse and humiliate our fellow human beings lingers in all of us, and the stories we pass on should serve as a reminder that none of us have achieved sainthood yet. How can we be sure how we would respond if pressured to take part in systematic cruelty organized and sponsored by those who are in power?

When we condemn the past and its evils, then, we may not make an exception for ourselves, however hard it may be to face up to our own sinfulness. The Christian practice of confessing one's sins — to God or to a priest or perhaps to a fellow believer — can help us here, and we see its wisdom mirrored in many a novel in which the protagonist has some ugly story about a terrible deed he committed long ago which he must get off his chest before he can round out his life. Clearly, honest confession is good for the soul. When I injected an element of moralizing into storytelling earlier (see Chapter 5 above), I did not exclude confession, for our narratizing about ourselves needs to be honest and historically truthful.

Jean-François Revel speaks of "two contradictory needs which are both equally strong and equally widespread: the need to justify one's existence at any price — and to declare oneself dissatisfied with it."[8] As we look back on our lives, we need to respond to both those needs. No wonder it is hard to get one's life story out: we are pulled in two separate directions at once!

Failed marriages. Among the people dissatisfied with their own past are many whose marriage has foundered. They must face the question: whose fault was it? All too often such a person decides it was his marriage partner's fault, and then he begins retelling the story

8. *On Proust*, tr. Martin Turnell (London: Hamish Hamilton, 1972), pp. 131–2. These two needs are mirrored in the opposite prayers uttered by the tax collector and the Pharisee (see Luke 18:9–14).

of his marriage in a way that is designed to absolve himself of blame. Unfortunately, the legal arrangements surrounding divorce encourage this kind of conduct, for in the battle over the assets and custody of the children, proven wrongdoing is a factor taken into consideration by the court.

The pattern to be discerned in narratives about failed marriages can also be found in stories told by young people when their romances break up during teenage years. The boyfriend who was nothing short of *wonderful* in last week's telling of the story may suddenly have turned into a "creep" after the relationship was broken off. Here we have the victor/victim situation again: if a young lady has broken off with a young man and now calls him a "creep," she is in some sense the winner in the contest which their relationship has become. And it is the way of many winners not to be generous in their storytelling; in fact, the actual break-up may even be arranged by her in such a way as to make for a "juicy" telling and retelling. In some cases the rejected boyfriend will fight back on a narrative level, but it is also possible that he will choose to suffer in silence. (Naturally, the genders could just as well have been reversed, and often are, when these things happen.)

What is especially striking is that narratives of failed marriages often produce an overall estimate of the marriage which is radically different from what was said in earlier years. A man may decide to leave his wife of twenty years, declaring that he is no longer happy with her, perhaps admitting in addition that he has found someone else. We may also find him insisting that he has *never* been happy with his wife — their marriage was unrelieved misery for him all along. Yet earlier he had often expressed himself as happy with the marriage. He may need the negative evaluation to prepare the way for a relationship with a new wife, who is insecure and needs to be told that she is superior to the former wife in every way. But if a second marriage follows upon the *death* of a first wife, there is room for both the husband and the second wife to be generous in talking about the first wife; somehow, there is no longer a need to triumph over her in narratives.

I have also seen such dynamics at work in organizations, where defeated candidates for positions and honors get rather rough treatment in retellings of what happened to their application or candidacy. It may be that a flaw or weakness that was discovered in the interview and evaluation process resulted in the candidate not getting the job — which is certainly fair enough. But as the story gets retold (by the person who *did* get the job, or perhaps by his allies), the flaw in the defeated candidate gets magnified unfairly and uncharitably, in a way that we deplore, just as we hate to see an obviously victorious team in football or basketball run up the score against its thoroughly demoralized opponents.[9] The danger of having one's opponents get carried away is one of the reasons why people are often reluctant to seek a higher position within an organization that already employs them, for they realize that the cost of *not* getting the position can be rather high: they may be in for a retrospective narrative mauling.

These regrettable tendencies are probably as common as are the more laudable instances in which we find people redeeming the past. If the past can be changed, this is surely good news insofar as it gives us some opportunity to right wrongs and undo damage we have done. But it can also be bad news, for it gives people the chance to do further damage to their enemies or to those they have already defeated. We should not regard such ugly behavior as inevitable or necessary, for it is certainly *possible* to be gracious in victory — whether in athletic competition or on the field of romance or in organizational life. Yet what we see all too often is a winner's vindictiveness as it manifests itself in a story that has come to be accepted as the truth of the matter.

Restitution. The enterprise of redeeming the past includes *action:* indeed, it is when we describe present and recent actions and also point to intended future actions, and then bring all of this into relation with something distressing that happened in the past, that we

---

9. Many coaches discourage their players from running up the score once victory is assured. Their reasons are quite sound and also apply to daily life outside competitive sports: an opponent who has been humiliated can become's a lifelong enemy. The Bible instructs us to *love* our enemies (see Matthew 5:43–8).

see what redeeming the past really comes to. The result is that a rather different narrative is able to emerge. A story that seemed to have a sad ending could wind up with a happy ending instead.

At the very least, restitution involves setting right what was done wrong. When a boy breaks a neighbor's window while playing baseball, he should offer to pay for the repair. Usually a broken window can be replaced in a fully satisfactory way, so that no trace of the damage is left. There are even repairs that strengthen or improve what was broken. The more difficult cases of restitution, however, involve damage or injury to human beings, and in some cases full restitution is utterly impossible. Yet restitution should be attempted even in such situations. The following narrative illustrates how a praiseworthy effort at restitution can be made in such cases.

> ◘ So you want to know how I happened to get involved in physiotherapy as a career? I guess it started with a car accident I was involved in: I ran over a thirteen-year-old boy riding his bicycle. He suffered serious spinal damage and wound up in a wheelchair — a quadriplegic. He'll be in that wheelchair for the rest of his life.
>
> The police who investigated the accident did not charge me with anything, nor did I get sued by the boy and his family. Yet I know that if I'd been driving more slowly and paying better attention, I could have avoided the accident. I've always felt guilty, even though friends tell me to forget it. They say, "It was an accident! You've got to put it behind you and get on with your life."
>
> Anyway, at first I couldn't bring myself to visit the boy I ran over — I just avoided thinking about him. But finally, one day I did visit him. And the visit seemed to mean a lot to him. It was then that I hit upon the idea of making physiotherapy my life's work — I suddenly wanted to work with people in *his* condition. So I went back to school, and I've never regretted the choice I made.
>
> No, I haven't ever considered offering my services as physiotherapist to the boy I ran over. Of course he's a man now — it happened years ago. I think it's better for him — and probably for me. It would be emotionally hard on both of us, I'm sure. But I still visit him once in a while, and he knows what work I do.

The Biblical injunction to repay evil with good (see I Thessalonians 5:15) can lead a person to cease a life of evil deeds and to devote himself to good instead: for example, a robber who repented and became a believer might work for a Christian agency offering agricultural assistance to people in Third-World countries. Surely such a career switch is praiseworthy, but there is something especially commendable about this case in which a man who is guilty of negligence (in his own eyes, at least) makes amends and redeems his past by devoting his life to undoing some of the consequences of similar negligence on the part of other people. In the process he takes an ugly episode from his own past, one which we might expect him to be reticent about, and makes it the virtual centerpiece of his life story, freely confessing his own guilt in the matter.

The effort to redeem the past in the spirit of making restitution should shape the concern that lives within people of good will. Because there is so much to be concerned about, we cannot help but seem blasé at times, freely turning away appeals for money and other assistance that come to us from unexpected quarters. We cannot get involved in *everything* or help *everyone.* Our own history may well help give direction to our concern for others and to our contribution to the wider community.

Yet that focused concern need not, as in the fictional case above, be rooted in guilt or wrongdoing. Sometimes the theme is gratitude more than restitution. It may be that someone had a mother who suffered from a certain chronic disease and needed a great deal of loving care and attention from doctors, nurses and community volunteers. That person might not go professionally into medical work but still retain a special interest in that disease and focus his concern and charitable giving and volunteer service on it. This, too, represents a commendable redemption of the past.

Too much continuity? In asking whether the concern for redeeming the past can lead to a quest for too much continuity, I am thinking especially of what we do about things that went wrong but did not have drastic consequences. The fiascoes of earlier years may well be grounded in sheer stupidity or short-sightedness rather than in any moral failing, although it is hard to separate these two elements as they come to expression in acts of wrongdoing. Is it necessary, in later life, to atone for every false start or grandiose project gone awry? The fact of the matter is that people are quite free when it comes to forgetting foolish episodes in which they were involved during their earlier years.

The desire to appear rational, which is much stronger in some people than in others, is a factor here.[10] Some people can laugh off the stupidities of youth quite freely, but others want to appear to have been following a single route all throughout their life, a route which they had in mind all along. As for those foolish episodes of yesteryear, either they get forgotten or it is shown, sometimes through rather circuitous reasoning, that they bore fruit in later successes.

The problem with the latter impulse (which I know well, since it is my own tendency) is that it quietly presupposes that everything somehow can or should be redeemed, that it should be taken along in the forward march of one's life. This presupposition embodies too high an estimate of our significance as (fallen) human beings. A more balanced analysis would recognize that a great deal of what we undertake does in fact turn out to be futile, like the wasted effort of the foolish man in Jesus' parable who built his house upon the sand (see Matthew 7:24–7).

Our day-to-day lives contain more incoherence than we generally admit. Even the sentences that issue from our mouths in daily conversation are quite disjointed; in the process of writing we get a chance to fix them up. Memory does quite some "fixing up" or editing for us as well: we tend to remember what worked, what led to success. Former acquaintances who were associated with us in some failing venture are easily forgotten.

In short, our sinful tendency toward self-righteousness and pride often corrupts the impulse to redeem the past. If we think in terms of *ourselves* alone, we are likely to go wrong. But if we remember that

10. The love of rationality (as opposed to the love of acting on impulse) is in good measure a product of our culture, with its long traditions of scientific and philosophical thought. But there are also elements in our world of culture and intellect that make people much more open to discontinuity and the new in their lives. Rationality and continuity are less valued today than they were during the 1950s.

our story is really part of something larger, that there is ultimately a *single story,* we may remember to put the emphasis where it belongs, namely, on God's sovereignty over our lives. As pilgrims on the earth, we cannot carry absolutely everything we have ever owned or handled along with us; rather, we are to "lay aside every weight, and sin which clings so closely" so that we can "run with perseverance the race that is set before us" (Hebrews 12:1).

Conversion and calling. The doctrine of God's sovereignty comes to expression in the Christian doctrine of conversion (especially its Calvinistic version). As something that *God* does or brings about, conversion surely represents a strong element of discontinuity in human life. We speak of people being turned right around, marching in the opposite direction, forsaking their sins, and so forth. But within Calvinistic theology there is also a covenantal emphasis, according to which God reaches out to his people and claims them as his own in a way that encourages them to remain part of the circle of fellowship and love into which they were born. On the one hand those who would follow Christ must leave their father and mother behind,[11] but on the other hand God places his seal upon us through baptism and leads and instructs us through our parents. Moreover, the process of salvation and sanctification is intended to restore us to what we were originally: salvation does not make its recipient into a new person who never existed before[12] but allows him to become who he really is. Our understanding of what it means to redeem the past must leave room for both these strands.

One outcome of the process of conversion is that we begin to rewrite the story of our past, this time leaving room for divine action, so that it is no longer just *our* story. Religious and political conversion are alike in this respect, for both open the prospect of "changing the past." Hayden White observes: "Such conversions may be seen as

---

11. See Matthew 12:46–50 and Luke 9:62: "No one who puts his hand to the plow and looks back is fit for the kingdom of God." The latter text reminds us of the calling of Elisha (see I Kings 19:19–21).

12. In some religious traditions, the convert takes on a new name, breaks his ties with the past, and even alters his appearance.

simply adopting a stance toward one's social and personal past that does not affect the historical past, the past as it really was, at all; but insofar as the new stance opens out a prospect that makes (or seems to make) a new kind of action possible in one's future, the historical past is affected as well." What happens when there is a *political* conversion of a certain sort is that ". . . events formerly regarded as unimportant are now redescribed as anticipations or prefigurations of the new society to be created by revolutionary action."[13]

Here we see that Christian thinking has a great deal in common with certain strands in nineteenth- and twentieth-century political thinking and philosophy of history; indeed, it inspired them in part. But in the Christian version of redeeming the past, there is less certainty about the new order to come. Instead the emphasis falls on God's sovereignty, and we realize ever more that we must overcome our tendency toward centeredness, our tendency to place ourselves right at the center of the action. The Christian doctrine of calling leads us toward a recognition that *God* assigns us our place in the drama that unfolds around us. We are to take the place he gives us and learn to redescribe our past in terms of that place. (Presumably he has prepared us for that role.) And we are to pray the petition that is at the very heart of the Lord's Prayer: "Thy will be done."

Participation in culture. God's calling and the place he assigns us seem to involve participation in culture. Some traditions within Christianity have stressed the necessity of cultural involvement, recognizing that there must be situations in which Christians work together closely with non-Christians. Other Christian groups have quietly acknowledged such involvement without making a problem of it. And then there are a few groups that have manifested considerable suspicion toward human culture — or, at least, toward some of its manifestations.[14] The so-called problem of Christianity and culture arises

---

13. "Getting Out of History: Jameson's Redemption of Narrative," in *The Content of the Form* (Chapter 5, Note 5), p. 150.

14. Christian thinkers, following the lead of H. Richard Niebuhr in *Christ and Culture*, usually delineate a number of distinct positions open to Christians. I have spelled out some non-Niebuhrian categories of my own in *Public Knowledge and*

largely because cultural endeavor involves the mingling of believers with non-believers.[15] Normally we do not feel any urge to philosophize about Christianity and culture when we see the Christian farmer alone in his field planting a crop. But much human activity is *cooperative* — it involves people who differ on various matters pooling their efforts.

The question to be faced here is what is implied in the widespread participation by Christians in cultural endeavors that are not under Christian control, or even of Christian inspiration or origin. Does such participation require us to regard culture in general, or culture taken as a whole, as involved in a process of being "redeemed" and incorporated into God's future for this world? The answer I would offer is that we should *not* simply say that it is redeemed; what we should affirm instead is that beginnings of all sorts — I am thinking of culture and technology as well as of social and even political action — are taken up into our continuing project and Christian future in such a way that when we redescribe the past, we can incorporate them as our own, as part of our past.

This answer is wholly within the spirit of Augustine's "treasures of the Egyptians" theme: part of Augustine's answer to the problem of Christianity and culture is that the people of God are to use the good things of this earth — the "treasures of the Egyptians,"[16] so to

---

Christian Education (Edwin Mellen Press, 1988), Chapters 5–6 and 9–10, and in *Christian Philosophy Within Biblical Bounds* (Neerlandia, Alberta: Inheritance Publications, 1990), Chapters 3–4 and 11–12.

15. Therefore the parable of the wheat and the tares is of great significance for the Christian assessment of cultural activity: see Matthew 13:24–30 and 36–43. I have commented on the broader significance of this passage in *Contending for the Faith: Heresy and Apologetics* (St. Catharines, Ont.: Paideia Press, 1984), pp. 81ff.

16. This phrase is an allusion to the passage in the book of Exodus where we read that when the Israelites finally left Egypt after the tenth plague, the people of Egypt gave them their treasures to take along. Unfortunately, the Israelites used those treasures to build the golden calf (see Exodus 32). Presumably they had a bit left over when it came time to build the tabernacle. See Augustine's *On Christian Doctrine,* tr. D. W. Robertson, Jr. (Indianapolis and New York: Bobbs-Merrill, 1958), Book II, Chapter 40, Section 60, p. 75; see also *Confessions,* Book VII, Section 9 (Chapter 1, Note 9), p. 146.

speak — for the service of God. Here Augustine echoes Psalm 105:37, where we read: "Then he led forth Israel with silver and gold." But the treasures of the Egyptians do not comprise the totality of human culture. Much of what man has created and produced is not fit for incorporation into the service of God and need not be taken up into the Christian redescription of the past that God makes possible for us. We must distinguish between treasures and trash.

One way to get at the continuing divisions between Christians on this set of questions is to ask whether believers should participate in the world's "dirty work." Here I am not just thinking of garbage collection but also of police work and military service[17] and of various occupations that might be deemed trivial or tedious or even comparatively useless. There are people who collect an income but provide no genuine or legitimate service, such as prostitutes. Some of the "work" done by lawyers also falls into this category.

Reformed Christians sometimes say that Christians should go out into "all areas of life," but this formulation, when taken literally, involves consequences that no Christian would want to affirm. What we wind up doing instead is participating in *some* of the dirty work, e.g. as garbage collectors, policemen, prison guards, but not all of it. As we consider this issue, we must not take the key term too literally. Dirty work is not just work that is dirty in a literal sense but also work that is unpleasant or harsh or performed under difficult circumstances. Although dentistry is an attractive and honorable occupation, it certainly includes elements of "dirty work," e.g. inflicting pain on people.

Some of the disagreements between Christians who discuss these matters have to do with the strategic question where we as Christians should concentrate our efforts and focus our resources as we engage in our daily work. Can the "dirty work" be left entirely to non-Christians? Some Christians say yes, while others maintain that this ap-

---

17. Part of the position on Christianity and culture held by many of the more traditional Anabaptists is that believers should not serve as policemen or soldiers. Should Christians who live in a country that retains capital punishment be willing to serve as executioner? Should they serve society by putting unwanted pets to death?

proach is unfair to the non-Christian part of the population.

The "dirty work" we do includes much that we will not take up into our redemptive redescriptions of the past.[18] If we accept a "dirty work" analysis as contributing to our understanding of the problem, we must be willing to admit that it is legitimate for Christians to be involved in work which may not bear much of the signature of God's coming kingdom, that is, work which may well be necessary in day-to-day terms (in the sense that *somebody* has to do it), but which is not of abiding significance.

It is God who assigns us our places in his kingdom and in the carrying out of his plans. To some (e.g. Martin Luther) he gives an exalted task; to others he assigns a humble (almost humiliating) role. And then there are some who, because of illness or some incapacity that dates from birth, are not able to contribute substantially to our material welfare: there seems to be nothing for them to do but pray, and perhaps to add warmth to the overall social and spiritual climate in the human community.

Alpha and omega. In this essay there has been movement back and forth between Christian thought, on the one hand, and issues arising in philosophy of history, on the other. There is a good reason for this interaction: Christian teaching has had a major influence on Western philosophical speculation about the nature of time and about the main lines of development in the historical process. But as Western thought has grown steadily more secular, it has also drifted away from a Christian sense of time and history. Therefore some of the points made in this essay will not fit in well with contemporary thought.

In particular, the notion that we are able to be citizens not just of the whole world (for the kingdom of God knows no borders) but of time far beyond the bounds of the razor-thin present will seem

18. Yet one well-known type of "dirty work" is widely sought by Christians, perhaps because it provides opportunities for spiritual contact with people, i.e. nursing. My convictions on the "dirty work" issue were shaped in part by my own experiences working as a nurse's aide while I was a college student.

strange. In Christian thought, the basis for this possibility lies in the doctrine of the image of God, which makes of man a creature who transcends the organic realm (which he shares with the animals). Because he bears God's image and is the object of God's electing love and has an eternal destiny awaiting him, man inhabits an expanded present and has the capacity to project himself back in thought to the very beginning (including the fall into sin on the part of Adam and Eve, which was also *our* fall) and ahead to the return of Christ and the consummation of the process of redemption and judgment.

Yet the emphasis should fall not on man's capacities to transcend his limited circumstances (wonderful as those capacities are in themselves) but on *God,* who gave them to man. It was God who called man into existence and gave him a special role and task, and it is God who guarantees that the promised goal will be reached, despite all the winding of the way, the false starts, and the times of utter despair when it appears that man may be his own undoing. God promises to "make all things new" (see Revelation 21:5), and in the light of that promise we see possibilities in the present and lines of continuity in the past which the eye of unbelief simply cannot see. *He* is the alpha and the omega, the beginning and the end, for whom not even the past is a closed book.

Glossary

The comments on terminology in this glossary overlap somewhat with passages in the text above. The glossary is intended as supplemental and has been provided for review and quick reference. Some readers may choose not to use it at all.

Anachronism. Ordinary dictionaries use this term in a sense that is less comprehensive than mine: an anachronism is something that does not belong in the time period in which it is placed. One dictionary I consulted gives the example of a building that is "out of harmony with its surroundings in point of time." In simple terms, an anachronism is an error regarding time or temporal order. I am using the term in a sense which allows anachronism to be a positive element in our thinking, for it liberates us from the confines of any one period of time.

Chronicle. A statement of what has happened in the past in which the element of narrative artistry is excluded or kept to an absolute minimum. A chronicle is written at about the time the events takes place and is not later altered in the light of other events.

Didacticism. The teaching of moral, or even practical, lessons through literature and stories. More broadly, a didactic approach to literature assumes that the purpose of each work is to teach or tell us something of a moral or philosophical or perhaps religious nature. The conception of narratizing which I defend leaves room for didactic elements.

Edited memories. This term is basically a metaphor, but a useful one. A text may be edited, cleaned up and regularized, which means that its parts are brought into line with each other, and probably with other texts as well; in similar fashion, we alter our memories, making them cohere with one another, with what we have learned from other people, and with our general outlook on life. The extent to which memories get edited differs from person to person.

Eschatology. The term used in theology for doctrines regarding the last things. Within Christian thought this term has to do with Christ's return and the events shortly before it and surrounding it. The term is also used more broadly to refer to expectations about the future which might come up in political philosophy or philosophy of history. Thus Marxism could be said to have an eschatology.

Hermeneutics. Interpretation theory, as applied especially to written texts, e.g. the Bible, more broadly to many forms of oral expression, and even to certain cultural artifacts, such as works of art. Hermeneutics enters into the consideration of narratives because of the interaction between the meaning of the part and the meaning of the whole on which hermeneutical theory tries to comment.

Historicism. Generally used to name any way of thinking that makes much of, and perhaps overestimates, the significance of historical considerations. The proper way to use this term is a matter of some scholarly dispute. Opponents of historicism claim that it leads to a debilitating relativism.

Literacy. See Orality.

Memory resources. They are not themselves memories but can be used to refresh and even reconstruct memories. For example, a victim of a criminal attack might have his memory of his ordeal strengthened and even clarified if he were brought back to the scene of the crime.

Memory supports. A narrower term than memory resources. Memory supports are items that serve to remind us of something or to help us remember, such as telephone directories, photographs of loved ones who have died, and history books. While they may date from the past, they are to be distinguished from traces of the past (see below) because they do not represent historical evidence in the direct sense. They are not normally — or properly — used to create memories but to buttress and sustain the memories we already have.

Mnemonic formula. "Mnémé" is a Greek word for memory. A mnemonic formula is a phrase that is easy to remember, such as a seven-digit telephone number with a hyphen between the third and fourth digits. Think also of such rhymed formulas as: "Thirty days hath September, / April, June, and November . . . ."

Omniscient narrator. Earlier novels often used an all-knowing narrator to tell the story; this narrator knew everything of relevance to the story and could be completely relied on in the account he offered. A great many narrators in more recent fiction know only in part (in this regard they are more like real human beings) and can be seen to slant their story along lines which their self-interest seems to demand. In some novels, two or more non-omniscient narrators cover the same ground.

Orality. A general word for reliance on speech (as opposed to language in its written form). The pre-school child who does not know how to read lives in an oral world, as do relatively primitive groups and tribes which do not have a way to write their language. Oral societies and tribes are heavily dependent on memory in the ordinary sense of that which we carry in consciousness and are able to recite when called upon. Societies that rely on written language are said to be literate or to possess literacy; they are less dependent on memory in the ordinary sense. Nevertheless, a literate society still possesses many traces and characteristics of orality.

Primary narrative. One that arises directly from memory and reflection on experience. When someone sees you looking frightened and asks what just happened to upset you, you normally respond with such a narrative. A primary narrative is to be distinguished from a refined and edited narrative, which arises from the process of revising your primary narrative when people ask you further questions or when new considerations arise in your own mind as you look back at the experience that gave rise to the narrative.

Prospective and retrospective discourse. When we project our awareness of present states of affairs into statements about the past, we are speaking retrospectively (looking backward), e.g. "I met my wife in a college math class" (she wasn't my wife on the day I met her) or "President Kennedy was born in 1917" (he didn't become the President until 1961). Prospective discourse looks ahead and describes the present in the light of anticipated successes or accomplishments, e.g. "I'm writing a Nobel-Prize-winning novel." (Whether the novel I am writing will win me the Nobel Prize remains to be seen; it is not even certain that I will get far enough with the project to claim that I wrote an unpublished novel.)

Reification. Something called into being (thing-ified) by language. If item X is a reification, we have a name for it and tend to assume that it exists, even if it does not correspond to any concrete particular in our experience: it may be no more than a thought. The justification of a reification might be that it helps us to order, arrange and classify the concrete particulars we encounter. To declare that there is "something" called memory is to engage in reification — or, at the very least, to accept memory as a concept or theoretical construct that has been posited by a process of reification.

Semanticized memory. The items contained in memory are not raw, unprocessed experiential givens; rather, they have been identified, interpreted, assigned a meaning, and brought into relation with other items that are remembered. The seemingly meaningless in human experience cannot be retained for long: either we assign it a meaning (semanticize it) or it fades away.

Traces of the past. Entities which were produced during a past era in which we have a historical interest, or which functioned in an important way during that era, and which can serve as a basis for our knowledge of it. It should be noted that in many cases such traces will have changed over the years. Thus there is more conjecture involved in basing knowledge on such traces than we often realize. We do not simply read them as texts that describe past events, although written records (sources) do fall under this heading as well. Traces of the past are what people usually refer to as historical evidence. Every form of historical evidence needs to be interpreted.

Index